ТЕОРИЯ СТРУКТУРНОГО АНАЛИЗА

TEORIYA STRUKTURNOGO ANALIZA

THE THEORY OF CRYSTAL STRUCTURE ANALYSIS

THE THEORY OF CRYSTAL STRUCTURE ANALYSIS

by

A. I. KITAIGORODSKII

Translated from the Russian
by David and Katherine Harker

CONSULTANTS BUREAU
NEW YORK
1961

The Russian text was published
by the USSR Academy of Sciences Press
in Moscow in 1957
under the editorship of B. K. Vainshtein

ISBN 978-1-4757-0342-9 ISBN 978-1-4757-0340-5 (eBook)
DOI 10.1007/978-1-4757-0340-5

Library of Congress Catalog Card Number: 60-8721

Softcover reprint of the hardcover 1st edition *1961*
227 West 17th St., New York 11, N. Y.

FOREWORD

Structure analysis is based on the phenomena of the diffraction of radiation by materials. In the first ten to twenty years after Laue's discovery, a very complete theory was developed for the diffraction of x-rays and, later, of electrons. This theory led to equations by means of which it was possible to compute the intensity pattern for a given structure. The theory of structure analysis came to mean that of the diffraction of radiation.

In 1935, Patterson pointed out a way leading to the solution of the inverse problem: the finding of the structure from a given intensity distribution pattern. At first the conservatism of researchers, and then the war, hampered the development and broad application of the ideas set forth in this work. It was only during the last ten years that all the rich possibilities of the Patterson method – the method of the analysis of the convolution of the electron density – were brought to light and applied in practice.

In 1947, Harker and Kasper found a new direct method for structure analysis – that of searching for relationships between structure amplitudes. In 1952 Sayre, Cochran, and Zachariasen published, in one and the same issue of Acta Crystallographica, interesting calculations and observations on the existence of an exceptionally simple relationship between structure amplitudes which permitted one to hope for the possibility of carrying out direct structure analyses by this method. Another new approach to the solution of the problem was that of Wilson, who showed a way of applying statistical considerations to the intensity distribution.

From 1947 to 1955, many authors added a great deal to the theory of direct methods of structure analysis.

Karle and Hauptman's researches, Goedkoop's short and elegant work, and that of many others were of substantial significance.

Thus, during the last five to ten years, a new realm of science was created – the theory of methods for finding structures, which took over at the point where activity in the theory of the diffraction of radiation stopped.

It is obvious at present that the specific problems of diffraction theory, for example, those of dynamic absorption, of anomalous scattering, and others, are of no importance in structure analysis (in the narrow sense of this concept). Therefore, identifying the theory of diffraction with that of structure analysis is unjust.

The term "theory of structure analysis"[1] is equivalent to a "theory of methods of finding the structure." It is to this theory that this book is devoted. This is a monograph, in the main generalizing the author's works of 1953-1955 directed toward the construction of a complete theory of the relationships between structure amplitudes. In his effort to give the book a character of completeness, the author attempted to touch upon all the questions in the theory of structure analysis that are being developed in the Soviet Union and abroad. Nevertheless, because of the particular character of the book, many valuable and interesting researches have been omitted.

The author hopes that the book will serve as an introduction to a new subject, and will serve as an impetus for further researches in this interesting realm, the goal of which is the expansion of the boundaries and the automatization of structure analysis.

[1]This term is the literal translation of the Russian title of this work. The word "crystal" has been added in the title of the translation to avoid ambiguity, but the exact translation of the term will be used throughout the text.

PUBLISHER'S NOTE

The vector notation used in the original Russian text has been retained in the translation. In this notation the scalar product is denoted by **ab** and the vector product by [**ab**]. The text also retains the symbols tg for tan and lg for log.

CONTENTS

CHAPTER I

MATHEMATICAL INTRODUCTION[1]

1. The Fourier Integral and Reciprocal Space

In practice, as is well known, every function satisfies the following formula put forward by Cauchy:

$$g(x) = \int_{-\infty}^{+\infty} e^{-2\pi i x\xi} d\xi \int_{-\infty}^{+\infty} g(x')\, e^{2\pi i x'\xi} dx'. \tag{1}$$

This formula can be extended to the case of three variables:

$$g(xyz) = \int_{-\infty}^{+\infty} e^{-2\pi i (x\xi + y\eta + z\zeta)} d\xi\, d\eta\, d\zeta \int_{-\infty}^{+\infty} g(x'y'z') e^{2\pi i (x'\xi + y'\eta + z'\zeta)} dx'\, dy'\, dz'. \tag{2}$$

Equations (1) and (2) are often referred to in the literature as Fourier double integrals.

From (2) it follows that

$$g(xyz) = \int_{-\infty}^{+\infty} G(\xi\eta\zeta)\, e^{-2\pi i (x\xi + y\eta + z\zeta)} d\xi\, d\eta\, d\zeta, \tag{3}$$

where

$$G(\xi\eta\zeta) = \int_{-\infty}^{+\infty} g(xyz)\, e^{2\pi i (x\xi + y\eta + z\zeta)} dx\, dy\, dz. \tag{4}$$

[1]This introduction is in the character of a summary. We limit ourselves to formal proofs only.

The functions g(xyz) and G($\xi\eta\zeta$) are called a pair of Fourier transforms.

Let xyz and $\xi\eta\zeta$ be the reference coordinates of two spaces.

The space of the xyz coordinates is defined by its metrics, that is by the three standard vectors $\mathbf{a}_i$. The vector **R** drawn from the origin of coordinates to the point x, y, z is equal to

$$\mathbf{R} = x\mathbf{a}_1 + y\mathbf{a}_2 + z\mathbf{a}_3.$$

The quantities xyz are relative numbers.

Furthermore, let the $\xi\eta\zeta$ space be defined by the metrics $\mathbf{b}_i$. The vector **H** drawn from the origin of coordinates of $\xi\eta\zeta$ space is equal to

$$\mathbf{H} = \xi\mathbf{b}_1 + \eta\mathbf{b}_2 + \zeta\mathbf{b}_3.$$

The quantities $\xi\eta\zeta$ are also relative numbers.

The volume element of xyz space is

$$dv = \mathbf{a}_i\,[\mathbf{a}_j\mathbf{a}_k]\,dx\,dy\,dz, \tag{5}$$

and that of $\xi\eta\zeta$ space is

$$d\tau = \mathbf{b}_i\,[\mathbf{b}_j\mathbf{b}_k]\,d\xi d\eta d\zeta. \tag{6}$$

Let the problem be to represent (2) in vector form. It is easy to see that this is impossible when there is no relationship between the metrics of the x- and ξ-spaces. Expression (2) can be written in vector form only if

$$\mathbf{a}_i\mathbf{b}_i = 1, \quad \mathbf{a}_i\mathbf{b}_k = 0. \tag{7}$$

From (7) it follows that

$$\mathbf{a}_i\,[\mathbf{a}_j\mathbf{a}_k] = \frac{1}{\mathbf{b}_i[\mathbf{b}_j\mathbf{b}_k]}. \tag{8}$$

It also follows from (7) that if the vectors $\mathbf{a}_i$ have the dimensions of length, then the vectors $\mathbf{b}_i$ have dimensions reciprocal to length. In the future, we shall call the xyz space physical, and the $\xi\eta\zeta$ space reciprocal.

With the aid of (7) and (8), we convert (2) into the form

$$g(\mathbf{R}) = \int e^{-2\pi i\mathbf{H}\mathbf{R}}d\tau \int g(\mathbf{R}')\,e^{2\pi i\mathbf{H}\mathbf{R}'}dv'. \tag{9}$$

Thus, a function of a vector in physical space can be represented by a Fourier double integral in vector form, on condition that there is a relationship between the metrics of the physical and reciprocal spaces given by (7).

2. Fourier Transforms

It follows from (9) that

$$g(\mathbf{R}) = \int G(\mathbf{H})\, e^{-2\pi i \mathbf{HR}} d\tau, \tag{10}$$

where

$$G(\mathbf{H}) = \int g(\mathbf{R})\, e^{2\pi i \mathbf{HR}} dv. \tag{11}$$

Functions g and G have different dimensions, so that $[G] = [g]\,[L^3]$.

We shall refer to the functions $g(\mathbf{R})$ defined in physical space and $G(\mathbf{H})$ in reciprocal space as a pair of Fourier transforms. To distinguish between (10) and (11) – insofar as the physical and reciprocal spaces are distinguishable – we shall refer to $G(\mathbf{H})$ as the Fourier transform of the function $g(\mathbf{R})$ and for brevity write it as

$$G = \Phi(g). \tag{11a}$$

We shall refer to the function $g(\mathbf{R})$ as the reciprocal Fourier transform of the function $g(\mathbf{H})$ and for brevity write it as

$$g = \Phi^{-1}(G). \tag{10a}$$

From (9), it is obvious that

$$G = \Phi\Phi^{-1}(G). \tag{12}$$

The same notation can be used for (3) and (4).

By using (10) and (11) it is easy to obtain the following relationship: if $G_1(\mathbf{H})$ and $G_2(\mathbf{H})$ are the Fourier transforms of the functions $g_1(\mathbf{R})$ and $g_2(\mathbf{R})$, then

$$\int G_1(\mathbf{H})\, G_2(\mathbf{H})\, d\tau = \int g_1(\mathbf{R})\, g_2(-\mathbf{R})\, dv. \tag{13}$$

Taking into account that

$$g^*(\mathbf{R}) = \int G^*(\mathbf{H})\, e^{2\pi i \mathbf{HR}} d\tau \tag{10b}$$

and

$$G^*(\mathbf{H}) = \int g^*(\mathbf{R})\, e^{-2\pi i \mathbf{HR}} dv \tag{11b}$$

(the complex conjugate of a function is marked with an asterisk), we also obtain, analogously to (13),

$$\int G_1(\mathbf{H})\, G_2^*(\mathbf{H})\, d\tau = \int g_1(\mathbf{R})\, g_2^*(\mathbf{R})\, dv. \tag{14}$$

An important special case of (14) is that of identical functions. Then

$$\int |G(\mathbf{H})|^2\, d\tau = \int |g(\mathbf{R})|^2\, dv. \tag{15}$$

3. Special Cases of Transforms

The triple integrals with which we worked in the preceding paragraph are reduced to single ones, either in the case of spherically symmetric functions, or when the function can be represented as a product in which each factor depends on one coordinate.

Let

$$G(\mathbf{H}) = G_1(\xi)\, G_2(\eta)\, G_3(\zeta).$$

Then (10) takes the form

$$g(\mathbf{R}) = \frac{1}{V} \int_{-\infty}^{+\infty} G_1(\xi)\, e^{-2\pi i x \xi} d\xi \int_{-\infty}^{+\infty} G_2(\eta)\, e^{-2\pi i y \eta} d\eta \int_{-\infty}^{+\infty} G_3(\zeta)\, e^{-2\pi i z \zeta} d\zeta, \tag{16}$$

that is

$$Vg(\mathbf{R}) = g_1(x)\, g_2(y)\, g_3(z), \tag{17}$$

where

$$g_i = \Phi^{-1}(G_i), \quad V = \mathbf{a}_i\, [\mathbf{a}_j \mathbf{a}_k].$$

Thus, in this case, transforms related to a function of one of the coordinates can be considered separately.

Let us assume now that $G(\mathbf{H})$ and, consequently, $g(\mathbf{R})$ have spherical symmetry. Introducing spherical coordinates, we write (10) as

$$g(R)=\int_0^{\infty}\int_0^{\pi}\int_0^{2\pi} G(H)\, e^{-2\pi i HR\cos\alpha} H^2 \sin\alpha\, dH\, d\alpha\, d\varphi. \qquad (18)$$

Integrating over the angles, we obtain

$$Rg(R)=2\int_0^{+\infty} HG(H)\sin 2\pi HR\, dH, \qquad (19)$$

that is, $Rg(R)$ and $HG(H)$ represent a pair of so-called Fourier sine transforms. Thus, if the spherically symmetrical $g(\mathbf{R})$ and $G(\mathbf{H})$ represent a pair of Fourier transforms, then $Rg(R)$ and $HG(H)$ are a pair of sine transforms.

Such simple interrelations between $g(\mathbf{R})$ and $G(\mathbf{H})$ do not arise in the case of cylindrical symmetry.

Let us give a few examples of transforms for the case of one variable.

Let $g(x) = e^{-\pi x^2}$. Let us find $\Phi(g)$. According to (4),

$$G(\xi)=\int_{-\infty}^{+\infty} e^{-\pi x^2} e^{2\pi i x\xi}\, dx.$$

Multiplying and dividing by $e^{-\pi\xi^2}$, we convert the integral to

$$\int_{-\infty}^{+\infty} e^{-\frac{1}{2}t^2}\, dt=\sqrt{2\pi}.$$

It follows that $G(\xi) = e^{-\pi\xi^2}$. From this it is apparent that the function and its transform have the same appearance. As a second example, let us take $g(x) = e^{-\beta|x|}$, $(\beta > 0)$. Then

$$G(\xi)=\int_0^{\infty} e^{-\beta x} e^{2\pi i \xi x}\, dx+\int_0^{\infty} e^{-\beta x} e^{-2\pi i \xi x}\, dx=\frac{2\beta}{\beta^2+4\pi^2\xi^2}\,.$$

Thus, $e^{-\beta|x|}$ and $\dfrac{2\beta}{\beta^2+4\pi^2\xi^2}$ are a pair of Fourier transforms.

For these functions (15) must be fulfilled. Actually,

$$2\int_0^{\infty} e^{-2\beta x}\, dx=\frac{1}{\beta}\,,$$

$$4\beta^2 \int_{-\infty}^{\infty} \frac{d\xi}{\beta^2 + 4\pi^2\xi^2} = \frac{1}{\beta}.$$

In the future we shall be interested almost exclusively in spherically symmetrical functions. Let us find the transforms of the functions

$$G_1 = Ze^{-\alpha H^2} \quad \text{and} \quad Ze^{-\alpha |H|}$$

with the aid of (19):

$$g_1 = \frac{2Z}{R} \int_0^{\infty} He^{-\alpha H^2} \sin 2\pi HR \, dH.$$

Integration gives

$$g_1(R) = Z \frac{\pi^{5/2}}{\alpha^{3/2}} e^{-\frac{\pi^2}{\alpha} R^2}; \tag{20}$$

$$g_2(R) = \frac{8\pi Z\alpha}{(\alpha^2 + 4\pi^2 R^2)^2}. \tag{21}$$

We are keeping in mind the difference in the form of the Fourier transforms of one-dimensional and three-dimensional functions (for example $e^{-\alpha x}$ and $e^{-\alpha R}$).

Finally let us examine the Fourier transform of a constant number, $g(x) = c$:

$$G(\xi) = c \int_{-\infty}^{+\infty} e^{2\pi i\xi x} dx. \tag{22}$$

In the future we shall use repeatedly the so-called δ function, determined by the following conditions: $\delta(x) = 0$ for all values of x except $x = 0$, and

$$\int_{-\infty}^{+\infty} \delta(x) \, dx = 1. \tag{23}$$

An important example of the δ function is the integral occurring in (22):

$$\int_{-\infty}^{+\infty} e^{2\pi i \xi x} dx = \delta(\xi). \tag{24}$$

Thus

$$G(\xi) = c\delta(\xi). \tag{25}$$

In this manner, a function having the same value at all points of space is the transform of a function which has the value zero in the whole of space with the exception of the origin of coordinates.

4. Projections and Sections of Transforms

Let g(**R**) and G(**H**) be a pair of Fourier transforms.

Let us examine the projection of g(**R**) on the plane A, the direction of the normal to which is **n**. Let us determine the projection of g(**R**) in the following manner:

$$g_A(\mathbf{R}) = \int_{-\infty}^{+\infty} g(\mathbf{R})\, dn, \tag{26}$$

where dn is the element of length in the **n** direction.

For convenience in computation, let us introduce a projectional system of coordinates $\mathbf{a}_i'$; furthermore let us direct one of these axial vectors (for example, $\mathbf{a}_3'$) along **n**. It is convenient to choose the coordinates $\mathbf{a}_1'$ and $\mathbf{a}_2'$ in such a way that they are the projection of the axial vectors $\mathbf{a}_1$ and $\mathbf{a}_2$ on the plane of the projection. Thus, in the projectional system of coordinates, $\mathbf{a}_3'$ is perpendicular to $\mathbf{a}_1'\mathbf{a}_2'$. Let us denote by $\mathbf{b}_i'$ the system of coordinates reciprocal to $\mathbf{a}_i'$.

In the new system of coordinates, (26) takes the form

$$g_A(x_1'x_2') = \int_{-\infty}^{+\infty} g(x_1'x_2'x_3')\, a_3' dx_3'. \tag{27}$$

We recall that the x_i are coordinates expressed in units of the axial vectors. Substituting

$$g(x_1'x_2'x_3') = \frac{1}{V} \int_{-\infty}^{+\infty} G(\xi_1'\xi_2'\xi_3')\, e^{-2\pi i (x_1'\xi_1' + x_2'\xi_2' + x_3'\xi_3')} d\xi_1' d\xi_2' d\xi_3', \tag{28}$$

we present (27) in the form

$$g_A(x_1'x_2') = \frac{1}{V}\int_{-\infty}^{+\infty} G(\xi_1'\xi_2'\xi_3')\, e^{-2\pi i(x_1'\xi_1'+x_2'\xi_2')} a_3 d\xi_1' d\xi_2' e^{-2\pi i x_3'\xi_3'} dx_3' d\xi_3'.$$

Taking advantage of the fact that, according to (24)

$$\int_{-\infty}^{+\infty} e^{-2\pi i x_3'\xi_3'} dx_3' = \delta(\xi_3'),$$

and using (23), we obtain

$$g_A(x_1'x_2') = \frac{1}{V}\int_{-\infty}^{+\infty} G(\xi_1'\xi_2'0)\, e^{-2\pi i(x_1'\xi_1'+x_2'\xi_2')} a_3' d\xi_1' d\xi_2'. \quad (29)$$

In this manner the following theorem is proved: if g (**R**) and G (**H**) are a pair of Fourier transforms, then the projection of one of the functions on the plane A represents the Fourier transform of the zero section by this plane of the other function (multiplied by the axial unit along the direction of projection):[2]

Let us now examine the projection of g (**R**) onto the **n** direction. Let us label it g_n(**R**).

In the same projectional system of coordinates g_n(**R**) will be defined in the following manner:

$$g_n(x_3') = \int_{-\infty}^{+\infty} g(R)\,|[\mathbf{a}_1'\mathbf{a}_2']|\, dx_1' dx_2'. \quad (30)$$

Substituting (28) and using the properties of the δ function, we obtain

$$g_n(x_3') = \frac{1}{V}\int_{-\infty}^{+\infty} G(00\xi_3')\, e^{-2\pi i x_3'\xi_3'}\,|[\mathbf{a}_1'\mathbf{a}_2']|\, d\xi_3'. \quad (31)$$

Thus, if g (**R**) and G (**H**) are a pair of transforms, then the projection of one of the functions onto the **n** direction represents the Fourier transform of

[2]In spite of the different dimensions of the direct and reciprocal spaces, we can combine them mentally.

the zero section by this direction of another function (multiplied by the area constructed on the vectors perpendicular to the direction of projection).

5. Convolution of Functions

We shall call the following integral the convolution S of two functions g_1 and g_2:

$$S(g_1, g_2) = \int g_1(\mathbf{R}') g_2(\mathbf{R} - \mathbf{R}') \, dv'. \tag{32}$$

The convolution of two functions given in physical space is, of course, a function of the coordinates of that same space. The indices 1 and 2 in the integral (32) can be interchanged:

$$S(g_1, g_2) = S(g_2, g_1). \tag{33}$$

The convolution of two functions of reciprocal space has exactly the same form:

$$S(G_1, G_2) = \int G_1(\mathbf{H}') G_2(\mathbf{H} - \mathbf{H}') \, d\tau'. \tag{32a}$$

Let us construct the Fourier transform of the product of two functions $g_1(\mathbf{R})$ and $g_2(\mathbf{R})$:

$$\Phi(g_1 g_2) = \int g_1(\mathbf{R}) g_2(\mathbf{R}) e^{2\pi i \mathbf{HR}} dv.$$

Using (10) for one of the functions under the integral, we obtain

$$\begin{aligned} \Phi(g_1 g_2) &= \int g_2(\mathbf{R}) e^{2\pi i \mathbf{HR}} dv \int G_1(\mathbf{H}') e^{-2\pi i \mathbf{H}'\mathbf{R}} d\tau' = \\ &= \int G_1(\mathbf{H}') \, d\tau' \int g_2(\mathbf{R}) e^{2\pi i \mathbf{R}(\mathbf{H}-\mathbf{H}')} dv = \\ &= \int G_1(\mathbf{H}') G_2(\mathbf{H} - \mathbf{H}') \, d\tau', \end{aligned} \tag{34}$$

that is, the Fourier transform of the product of two functions is equal to the convolution of the Fourier transforms of these functions:

$$\Phi(g_1 g_2) = S(G_1, G_2). \tag{35}$$

In other words, g_1g_2 and $S(G_1, G_2)$ represent a pair of Fourier transforms. Therefore, the inverse Fourier transform of the convolution of the functions G_1 and G_2 is equal to their product

$$\Phi^{-1}[S(G_1, G_2)] = g_1g_2. \quad (36)$$

The last equation can be written

$$\Phi^{-1}[S(G_1, G_2)] = \Phi^{-1}(G_1)\Phi^{-1}(G_2). \quad (37)$$

Reformulations analogous to (34) carried out for the product of the functions G_1G_2 give

$$\Phi^{-1}(G_1G_2) = S(g_1, g_2); \quad (35a)$$

$$\Phi[S(g_1, g_2)] = G_1G_2; \quad (36a)$$

$$\Phi[S(g_1, g_2)] = \Phi(g_1)\Phi(g_2). \quad (37a)$$

Thus, if in physical (reciprocal) space one function is the product of two others, then in reciprocal (physical) space the transform of this function is the convolution of the transforms of the two others.

The process of forming the convolution can obviously be repeated. The order in which the convolutions are carried out is of no importance.

The commutativity of the convolution operation follows from (35). Let us illustrate this important property of the convolution by the example of four functions. The convolution of four functions can be carried out either by convoluting one pair of functions, then the other, and finally by convoluting the two convolutions:

$$S[S(g_1, g_2), S(g_3, g_4)], \quad (38)$$

or by convoluting two functions, then convoluting the third with the convolution of the two, and, finally, convoluting the fourth with the convolution of the three functions:

$$S\{g_1, S[g_2, S(g_3, g_4)]\}. \quad (39)$$

With the aid of (35a) let us convert (38) first to

$$S[\Phi^{-1}(G_1G_2), \Phi^{-1}(G_3G_4)]$$

then, using the same equation for the second time, and also taking account of (12), we obtain

$$\Phi^{-1}(G_1G_2G_3G_4) = S\,[S(g_1,\ g_2),\ S(g_3,\ g_4)]. \tag{40}$$

It is quite obvious that a permutation of indices does not alter the convolution (40), since the left side of the equation contains the product of the functions G_i, which is commutative.

In exactly the same manner, applying (35a) repeatedly, we obtain from (39)

$$S\{g_1,\ S\,[g_2,\ \Phi^{-1}(G_3G_4)]\} = S\,[g_1,\ \Phi^{-1}(G_2G_3G_4)] = \\ = \Phi^{-1}(G_1G_2G_3G_4).$$

Thus, the convolution of the functions in any sequence leads to the same results. It is possible to speak of an n function convolution as a quite definite function, without indicating the sequence of convolutions.

In integral form, the convolution of n functions taken in any sequence can always be reduced to the following form

$$\begin{aligned} &S(g_1,\ S(g_2,\ S\ldots\ 1,\ S(g_{n-1},\ g_n))) = \\ &= \int g_1(\mathbf{R}_1)\, g_2(\mathbf{R}_2)\, g_3(\mathbf{R}_3) \ldots g_n(\mathbf{R} - \mathbf{R}_1 - \\ &- \mathbf{R}_2 - \ldots - \mathbf{R}_{n-1})\, dv_1 dv_2 \ldots dv_{n-1}. \end{aligned} \tag{41}$$

The convolution of n functions is expressed by an (n − 1)-fold multiple integral.

Let us again illustrate (41) by the example of four functions:

$$S(g_1,\ g_2) = \int g_1(\mathbf{R}_1)\, g_2(\mathbf{R} - \mathbf{R}_1)\, dv_1;$$

$$S(g_3,\ g_4) = \int g_3(\mathbf{R}_3)\, g_4(\mathbf{R} - \mathbf{R}_3)\, dv_3;$$

$$S\,[S(g_1,\ g_2),\ S(g_3,\ g_4)] =$$

$$= \int dv_1 \int dv_3 \int g_1(\mathbf{R}_1)\, g_2(\mathbf{R}'_2 - \mathbf{R}_1)\, g_3(\mathbf{R}_3)\, g_4(\mathbf{R} - \mathbf{R}'_2 - \mathbf{R}_3)\, dv'_2.$$

Substituting the variable $\mathbf{R}'_2 - \mathbf{R}_1 = \mathbf{R}_2$, we obtain (41).

For brevity we use vectors. If g_1 and g_2 are expressed as functions of relative coordinates, then (32) appears as

$$S(g_1,\ g_2) = \int_{-\infty}^{+\infty} g_1(x'y'z')\, g_2(x - x',\ y - y',\ z - z')\, dv'.$$

In the case of the one-dimensional problem the convolution of two functions has the form

$$S(g_1, g_2) = \int_{-\infty}^{+\infty} g_1(x') g_2(x - x')\, dx'. \tag{32b}$$

What then is the meaning of this integral? Figure 1 shows the two functions $g_1(x')$ and $g_2(x')$. The function $g_2(x - x')$ has the following meaning: it is a $g_2(x')$ function inverted (in the one-dimensional case, one can say reflected) at the origin of coordinates and shifted by x in a positive direction. The shaded area is the value of the convolution for the argument x. The curve enclosing this area is obtained by multiplying $g_2(x - x')$ by $g_1(x')$.

Let us examine the simplest examples of convolutions.

Let $g_1(x') = A$ between the limits a and b, and $g_1(x') = 0$ when $x' < a$ and $x' > b$. Then

$$S(g_1, g_2) = A \int_a^b g_2(x - x')\, dx' = A \int_{x-b}^{x-a} g_2(t)\, dt. \tag{42}$$

In spite of the fact that one of the functions has a constant value, the convolution is not at all similar to the second function. Such similarity will occur only when one of the functions is a δ function. Actually, when $g_1(x') = \delta(x')$, (32b) gives

$$S(\delta, g_2) = g_2(x). \tag{42a}$$

The case when one of the functions is a sum of δ functions is very interesting and important for what is to come.

Thus, if $g_1(x') = \delta(x' - a) + \delta(x' - b) + \delta(x' - c)$, then

$$S(g_1, g_2) = g_2(x - a) + g_2(x - b) + g_2(x - c). \tag{43}$$

In this manner the convolution of a function by a sum of δ functions propagates the second function, as is shown in Fig. 2.

Let us now answer the following question. If g_1 differs from zero only within the limits a and b, and g_2 differs from zero only from α to β, then in what region does their convolution differ from zero?

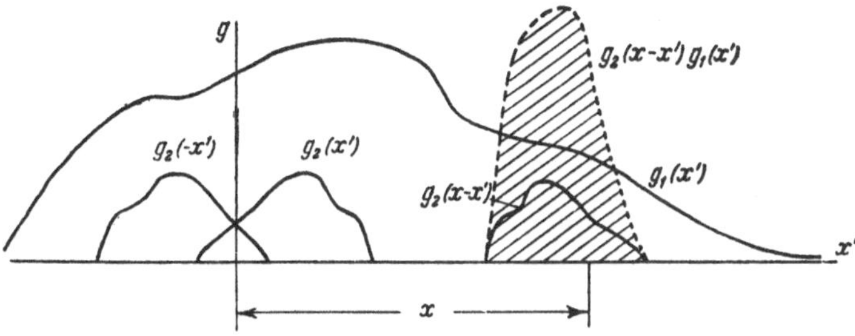

Fig. 1. Convolution of the functions $g_1(x)$ and $g_2(x)$.

$$S(g_1, g_2) = \int_{-\infty}^{+\infty} g_1(x') g_2(x-x')\,dx' = \int_{-\infty}^{a} \cdots + \int_{a}^{b} \cdots + \int_{b}^{+\infty} \cdots$$

The two extreme integrals fall away because of the limitation imposed on the first function:

$$S(g_1, g_2) = \int_{a}^{b} g_1(x') g_2(x-x')\,dx'.$$

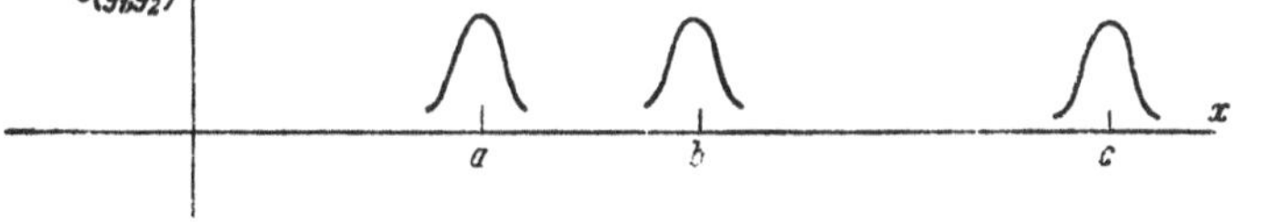

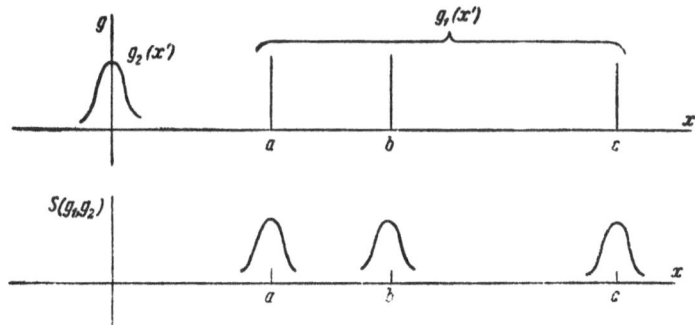

Fig. 2. Convolution of an arbitrary function with a sum of δ functions.

Let us make a substitution of the variable $t = x - x'$. Then

$$S(g_1, g_2) = \int_{x-b}^{x-a} g_1(x-t) g_2(t)\,dt.$$

Now we take into account the limited interval in which the second function differs from zero. The lower limit must not exceed β, the upper must not be smaller than α. Consequently, the convolution differs from zero when

$$a+\alpha < x < b+\beta. \tag{44}$$

Therefore, the limits of the interval in which the convolution differs from zero are equal to the sum of the limits of the intervals in which the convoluted functions differ from zero. This interval may be taken to be outside the intervals from a to b and from α to β.

Let us now calculate the convolution of two Gaussian functions:

$$g_1(x) = \frac{h_1}{\sqrt{\pi}} e^{-h_1^2 x^2} \text{ and } g_2(x) = \frac{h_2}{\sqrt{\pi}} e^{-h_2^2 x^2}.$$

The convolution is equal to

$$S(g_1, g_2) = \frac{h_1 h_2}{\pi} \int_{-\infty}^{+\infty} e^{-h_1^2 x'^2 - h_2^2 (x-x')^2} dx'.$$

Representing the exponent in the form

$$-(h_1^2+h_2^2)x'^2 + 2h_2^2 x x' - h_2^2 x^2 =$$

$$= -(h_1^2+h_2^2)\left(x' - \frac{h_2^2 x}{h_1^2+h_2^2}\right)^2 - \frac{h_1^2 h_2^2}{h_1^2+h_2^2} x^2$$

and introducing the variable of integration

$$x' = \frac{h_2^2 x}{h_1^2+h_2^2},$$

we obtain

$$S(g_1, g_2) = \frac{H}{\sqrt{\pi}} e^{-H^2 x^2}, \tag{45}$$

where

$$H = \frac{h_1 h_2}{\sqrt{h_1^2+h_2^2}}.$$

In this manner, it turns out that the convolution of two Gaussian functions is also a Gaussian function, but with a different dispersion.

Attention should be given to the meaning of the convolution of two functions in the theory of probability.

Let t_1 and t_2 be two independent random quantities, and $x = t_1 + t_2$ be their sum.

Let us find the probability $\varphi(x)$ that the sum $t_1 + t_2$ is smaller than the value of $\underline{x}$. Let $f(t_1)dt_1$ be the probability that t_1 lies in the interval from t to t_1; and let us make similar assumptions for t_2. Then

$$\varphi(x) = \iint f_1(t_1) f_2(t_2)\, dt_1 dt_2,$$

where the integral must be taken over the unbounded half plane limited by the straight line $t_1 + t_2 = x$, that is;

$$\varphi(x) = \int_{-\infty}^{+\infty} \int_{-\infty}^{x-t_1} f_1(t_1) f_2(t_2)\, dt_1 dt_2.$$

The probability density of the sum $t_1 + t_2 = x$ is equal to

$$\varphi'(x) = \int_{-\infty}^{+\infty} f_1(t) f_2(x-t)\, dt, \tag{46}$$

and this is nothing but $S(f_1, f_2)$. Thus, the probability density of the sum of two independent magnitudes is the convolution of the probability densities of the summands.

6. Self-Convolutions of Functions

In agreement with the terminology established in the mathematical literature, we determined in Section 5 the convolution of functions $g_1(\mathbf{R})$ and $g_2(\mathbf{R})$ as

$$S(g_1, g_2) = \int g_1(\mathbf{R}') g_2(\mathbf{R} - \mathbf{R}')\, dv' \tag{32}$$

and showed [see (34)-(37)] that the transform of the convolution is equal to the product $G_1(\mathbf{H})\, G_2(\mathbf{H})$ of the Fourier transforms of the convoluted functions. It turned out that (see Fig. 1) the integrand is the product of one function by the second function translated and inverted at the origin of coordinates.

Let us form the integral

$$\int g_1(\mathbf{R}') g_2(\mathbf{R}+\mathbf{R}')\, dv', \tag{47}$$

which is not the convolution of the functions $g_1(\mathbf{R})$ and $g_2(\mathbf{R})$. Its integrand is the product of one function by a second function translated (but not inverted) with respect to the origin of coordinates. Let us emphasize that (47) is not equal to (32).

Writing (47) in the form (the variable of integration $\mathbf{R}'$ can be substituted by $-\mathbf{R}'$)

$$\int g_1(-\mathbf{R}') g_2(\mathbf{R}-\mathbf{R}')\, dv',$$

we find that (47) is exactly $S(g_1^i, g_2)$, where

$$g_1^i(\mathbf{R}) = g_1(-\mathbf{R}). \tag{48}$$

If $g_1(\mathbf{R})$ is centrosymmetric, then (47) and (32) are identical.

Integral (47) is the convolution of two functions of which one has first been inverted.

$$S(g_1^i, g_2) = \int g_1(\mathbf{R}') g_2(\mathbf{R}+\mathbf{R}')\, dv'. \tag{49}$$

The Fourier transform of the convolution $S(g_1, g_2)$ is the product of the Fourier transforms of the functions g_1 and g_2. Consequently,

$$\Phi S(g_1^i, g_2) = \Phi(g_1^i)\Phi(g_2). \tag{50}$$

In the general case of an arbitrary complex function $g_1(\mathbf{R})$, it is impossible to pass from $\Phi(g_1^i)$ to $\Phi(g_1) = G_1$.

The situation changes if $g_1(\mathbf{R})$ is a real function. Then $g_1(\mathbf{R}) = g_1^*(\mathbf{R})$ and, consequently, according to (11b),

$$\begin{aligned}\Phi(g_1^i) &= \int g_1^i(\mathbf{R})\, e^{2\pi i \mathbf{HR}} dv = \int g_1(-\mathbf{R})\, e^{2\pi i \mathbf{HR}} dv = \\ &= \int g_1(\mathbf{R})\, e^{-2\pi i \mathbf{HR}}\, dv = \int g_1^*(\mathbf{R})\, e^{-2\pi i \mathbf{HR}}\, dv = G_1^*(\mathbf{H}).\end{aligned} \tag{51}$$

Thus, if $g_1(\mathbf{R})$ is real, then

$$\Phi \int g_1(\mathbf{R}') g_2(\mathbf{R}+\mathbf{R}')\, dv' = G_1^*(\mathbf{H})\, G_2(\mathbf{H}). \tag{52}$$

Let us now introduce the concept of the self-convolution.

We shall call (49) the self-convolution [1,2] of a function $g(\mathbf{R})$ for the case where the convoluted functions are identical:

$$S(g^i,\ g) = \int g(\mathbf{R}') g(\mathbf{R}+\mathbf{R}')\, dv'. \tag{53}$$

Substituting the variable $\mathbf{R}+\mathbf{R}' = \mathbf{R}''$, we obtain

$$S(g^i,\ g) = \int g(\mathbf{R}''-\mathbf{R})\, g(\mathbf{R}'')\, dv''. \tag{53a}$$

Comparing (53a) with the preceding formula, we come to the conclusion that the self-convolution of a function has a center of symmetry

$$S^i(g^i,\ g) = S(g^i,\ g). \tag{54}$$

The transform of the self-convolution is given by (50). However, for a real $g(\mathbf{R})$ we have

$$\Phi \int g(\mathbf{R}') g(\mathbf{R}+\mathbf{R}')\, dv' = G(\mathbf{H})\, G^*(\mathbf{H}) = |G^2(\mathbf{H})|. \tag{55}$$

The transform of the self-convolution of a real function is a real positive centrosymmetric function.

It follows from (55) that for a real g

$$S(g^i,\ g) = \Phi^{-1}(|G^2|) = \int |G^2(\mathbf{H})|\, e^{-2\pi i \mathbf{HR}}\, d\tau. \tag{56}$$

Since the self-convolution is centrosymmetric, (56) becomes

$$S(g^i,\ g) = 2\int G^2(\mathbf{H}) \cos 2\pi \mathbf{HR}\, d\tau. \tag{57}$$

Let us emphasize that self-convolution is not a special case of convolution. Thus,

$$S(g,g) = \int g(\mathbf{R}')\, g(\mathbf{R}-\mathbf{R}')\, dv', \tag{58}$$

which is not the same as (53). However, in one important case (53) and (58) coincide, i.e., when the function g(**R**) has a center of symmetry or of antisymmetry. Only then

$$S(g, g) = S(g', g). \tag{59}$$

CHAPTER II

PRINCIPLES OF THE THEORY

1. Scattering of X-Rays

X-rays are scattered by electrons. Therefore the electron distribution in a body can be found by the method of x-ray structure analysis.

Let us choose in an object an arbitrary origin of coordinates, and symbolize by $\rho(\mathbf{R}, t)\,dv$ the number of electrons present in the element of volume dv at a distance **R** from this origin at the instant t. It only makes sense to speak of those changes in $\rho(\mathbf{R}, t)$ with time which are determined by the thermal vibrations of the atoms and not by the motions of the electrons within the atom.

The radiation scattered by the object is divided into coherent and incoherent. The latter is accompanied by a change in wavelength.

Experiment allows us to establish the dependence of the intensity on the scattering angle for various arrangements of the object in relation to the impinging ray. The incoherent scattering does not change with a change in the orientation of the object, since its intensity is not connected with the structure of the object, but is determined only by the atomic numbers of the atoms of which the body is composed.

The change of the incoherent scattering with the scattering angle can be computed from formulas derived by Compton, Waller, Heisenberg, and others. In the practice of x-ray structure analysis the following formula is widely used (for an atom with atomic number Z): the intensity of the incoherent radiation is

$$I = \sum_{j=1}^{Z} (1 - |f_j|^2),$$

where f_j is the atomic factor of the atom with the number j (see, e.g., [3] ch. 7).

In an overwhelming majority of cases, the part which is coherent and subject to classical electrodynamics can easily be separated experimentally from the general scattering. Only this basic part of the scattering is of interest in structure analysis.

The appearance of scattering, according to classical electrodynamics, occurs because the impinging wave causes the electrons to vibrate with the frequency of the radiation from without. Each electron therefore becomes the source of a secondary wave.

The phase of the wave at the given point **R** (considering the phase at the origin of coordinates to be zero) is equal to **kR**, where **k** is the wave vector having the direction of the wave front and the magnitude $|\mathbf{k}| = \frac{2\pi}{\lambda}$.

Let the wave fall on the element dv, situated at point **R**, in the direction $\mathbf{k}_0$ and be scattered in the direction **k**. Then the phase of the wave leaving dv will be $\mathbf{kR} - \mathbf{k}_0\mathbf{R}$. The amplitude of the scattered wave will be proportional to the number of electrons present in dv. The expression for the amplitude of the wave scattered at a given point of space is proportional to

$$\rho(\mathbf{R},\ t)\, e^{i(\mathbf{k}-\mathbf{k}_0)\mathbf{R}}\, dv. \tag{1}$$

Let us introduce the vector

$$\mathbf{H} = \frac{1}{2\pi}(\mathbf{k} - \mathbf{k}_0), \tag{2}$$

which has dimensions reciprocal to length. The vector **H** can be represented in reciprocal space by formula (I, 4).[1]

It is not hard to see that

$$H = \frac{2\sin\vartheta}{\lambda}, \tag{3}$$

where ϑ is half of the scattering angle, i.e., half of the angle between the impinging and the scattered rays. Thus, the volume element of the body scatters at the moment[2] t an x-ray with an amplitude proportional to

[1]In referring to formulas from other chapters the first figure denotes the number of the chapter.

[2]One should remember that the period of vibration of the x-ray wave is incomparably smaller than that of the thermal vibrations.

$$\rho(\mathbf{R},\ t)\, e^{i2\pi\,\mathbf{HR}}\, dv. \tag{4}$$

If the action of the wave scattered by one electron on another electron is not taken into account, then the amplitude of the wave scattered by the object is the algebraic sum of the amplitudes of the waves scattered by each of the electrons:

$$G(\mathbf{H},\ t) = \int \rho(\mathbf{R},\ t)\, e^{i2\pi\mathbf{HR}}\, dv\,. \tag{5}$$

The rough assumption of the absence of interaction of elementary waves, made by Laue in constructing the first (so-called kinematic) theory of x-ray diffraction, is well confirmed by experiment. The reason for this good agreement is that a real crystal is broken into small mosaic blocks. Inside a small block the action of the scattered waves on the other centers of scattering is negligible. Formula (5) applies to just such a block. Since the blocks are displaced in relation to each other, the resulting scattering is obtained through the addition, not of amplitudes, but of the scattering intensities of the blocks (coherence exists only within a block). Therefore, (5) is also correct for the whole object to the accuracy needed in practice (we shall see below that a knowledge of the amplitudes to the order of accuracy of 10% is quite sufficient for structure analysis).

The square of the scattering amplitude of the whole object for a given value of **H**, when the amplitude is represented by a complex number, is equal, as is well known, to the product of the amplitude by its complex conjugate, or, which is the same thing, to the square of the modulus of the complex amplitude,[3] i.e.,

$$|G(\mathbf{H},\ t)|^2 = GG^* = \iint \rho(\mathbf{R},\ t)\, \rho(\mathbf{R}',\ t)\, e^{i2\pi\mathbf{HR}}\, e^{-i2\pi\mathbf{HR}'}\, dv\, dv'. \tag{6}$$

[3] The wave equation in complex form appears as $Ge^{i\omega t}$, in the real form it appears as $A\cos(\omega t + \varphi)$. Between these two representations there is the relationship

$$Re\,\{Ge^{i\omega t}\} = A\cos(\omega t + \varphi).$$

Since G is a complex magnitude, the left side of the equation has the form $G_{real}\cos\omega t - G_{imag}\sin\omega t$. The right side of this equation can be written as: $A\cos\varphi\cos\omega t - A\sin\varphi\sin\omega t$. Consequently, $G_{real} = A\cos\varphi$, $G_{imag} = A\sin\varphi$. From this $A^2 = (G_{real})^2 + (G_{imag})^2 = GG^*$.

The mean square of the amplitude of the wave (when speaking of the square of the amplitude, we always have in view the square of the modulus) scattered for a given value of **H** defined in (6), determines the scattering intensity measured experimentally, namely

$$I(\mathbf{H}) = \varphi(\mathbf{H})\,\overline{G^2(\mathbf{H},\ t)}. \tag{7}$$

The averaging must be carried out over time, since the duration of the measurement is much greater than the period of the thermal vibrations.

To convert the square of the amplitude into intensity, it is necessary, not only to average G^2, but also to multiply it by some function of **H** which depends on the conditions of the experiment. The necessity of such a multiplier is clear from the following considerations: the finite size of the object, and the inevitable divergence of the initial and secondary beams, waves for which the values of the **H** vector are distributed over a certain interval $\mathbf{H} \pm \Delta\mathbf{H}$ superimposed at one point of observation.

Consequently, to compute the scattering intensities in this average **H** direction, it is necessary to integrate (6) over the region $\Delta\tau$. The result of such an integration can always be represented in the form of (7) (theorem of the mean).

If G^2 is a slowly varying function of **H** (in the case of gases, liquids, or amorphous bodies) then $\varphi(\mathbf{H})$ = const. In crystals G^2 changes very rapidly with **H** and for them the function $\varphi(\mathbf{H})$ is computed for the various methods of measuring intensity ([4] ch. III, sec. 6).

The existing methods of analyzing crystals are such that $\varphi(\mathbf{H}) = \varphi(H)$, i.e., the conversion coefficient depends only on the magnitude of **H**, but not on its direction.

The structure of the object is determined by the scattering amplitude (5). With the aid of (7), experiment allows one to find the value of the time average of the square of the scattering amplitude. The finding of the structure from the values of $G^2(\mathbf{H},t)$ is the basic problem of structure analysis.

We shall call the magnitude $\overline{G^2(\mathbf{H},t)}$ the induced intensity of scattering or, if it does not lead to misunderstanding, simply the intensity of scattering.

2. Reciprocal Space as the Space of Scattering Functions

It follows from the preceding that the amplitude and intensity of scattering can be defined as functions of position in reciprocal space. Amplitudes and intensities of scattering are functions of the vector **H**, the length of which is determined by the magnitude of the scattering angle according to (3), while its direction depends on the orientation of the object with respect to the ray at the instant of scattering, according to (2).

Thus, the values of the scattering intensity can be mentally laid off in an imaginary three-dimensional space, either as numbers at each point in space, or in the form of compressions and rarefications of a kind of cloud. If the intensity differs from zero only at certain values of **H**, then it means that reciprocal space is not completely filled, and that there are isolated regions where the function differs from zero. As we shall see below, we shall come across intensity distributions for which reciprocal space presents a system of points, a system of rods, a system of layers, etc.

In order to construct functions of the intensity or the scattering amplitude it is possible in general to choose any system of coordinates. However, a series of new possibilities arises if we connect the system of coordinates of the object with that of reciprocal space by the conditions (I, 7).

If this is done, then (5), the fundamental formula of the theory, acquires the following meaning

$$G(\mathbf{H},\ t) = \Phi[\rho(\mathbf{R},\ t)]. \tag{8}$$

Thus if the amplitude of scattering is represented as a function of the vector **H**, determined by (2), and if the coordinate systems of reciprocal space and the space of the scattering object obey (I, 7), then the scattering amplitude is the Fourier transform of the electron density of the scattering object.

The terms reciprocal and Fourier space can be used interchangeably when the relationships (I, 7) apply.

3. The Time Average of the Electron Density

In the preceding paragraph, it was stressed that experiment allows one to carry out a measurement only of the time average value of the intensity $G^2(\mathbf{H},t)$. It is obvious, therefore, that (8) has no direct practical application. There is also no doubt that the measurements do not allow one to determine directly the instantaneous values of $\rho(\mathbf{R},t)$.

It is therefore essential to find, instead of (8), equations which would help to furnish important information about the structure from the time average, and not the instantaneous values, of the amplitudes and intensities of x-ray scattering.

Let us symbolize the time average of the electron density of the object by $\bar{\rho}(\mathbf{R})$

$$\overline{\rho(\mathbf{R},\ t)}=\bar{\rho}(\mathbf{R}). \tag{9}$$

Then

$$\rho(\mathbf{R},\ t)=\bar{\rho}(\mathbf{R})+\Delta_T(\mathbf{R},\ t), \tag{10}$$

in which the time average of the last term is

$$\overline{\Delta_T(\mathbf{R},\ t)}=0. \tag{11}$$

Averaging over time the expression for the intensity of scattered radiation (6), it is possible with the aid of (10) to obtain the following:

$$\begin{aligned} G^2(\mathbf{H},\ t) = {} & \iint \bar{\rho}(\mathbf{R})\,\bar{\rho}(\mathbf{R}')\,e^{i2\pi\mathbf{HR}}\,e^{-i2\pi\mathbf{HR}'}\,dv\,dv' + \\ & + 2\iint \bar{\rho}(\mathbf{R})\,\Delta_T(\mathbf{R}',\ t)\,e^{i2\pi\mathbf{HR}}\,e^{-i2\pi\mathbf{HR}'}\,dv\,dv' + \\ & + \iint \Delta_T(\mathbf{R},\ t)\,\Delta_T(\mathbf{R}',\ t)\,e^{i2\pi\mathbf{HR}}\,e^{-i2\pi\mathbf{HR}'}\,dv\,dv'. \end{aligned} \tag{7}$$

A line above a quantity means that it is to be averaged over time. Changing the order of averaging (i.e., integration over time and integration over volume) and taking (11) into account, we obtain for the time average intensity the expression

$$\overline{G^2(\mathbf{H},\ t)}=\mathfrak{F}^2(\mathbf{H})+\mathfrak{F}_T^2(\mathbf{H}), \tag{12}$$

where

$$\mathfrak{F}^2(\mathbf{H})=\iint \bar{\rho}(\mathbf{R})\,\bar{\rho}(\mathbf{R}')\,e^{i2\pi\mathbf{HR}}\,e^{-i2\pi\mathbf{HR}'}\,dv\,dv', \tag{13}$$

$$\mathfrak{F}_T^2(\mathbf{H})=\iint \overline{\Delta_T(\mathbf{R},\ t)\,\Delta_T(\mathbf{R}',\ t)}\,e^{i2\pi\mathbf{HR}}\,e^{-i2\pi\mathbf{HR}'}\,dv\,dv'. \tag{14}$$

If the vibrations of the electron density at different points of the object occur independently of one another, then

$$\overline{\Delta_T(\mathbf{R},\ t)\,\Delta_T(\mathbf{R}',\ t)}=\overline{\Delta_T(\mathbf{R},\ t)}\cdot\overline{\Delta_T(\mathbf{R}',\ t)}=0, \tag{15}$$

and the integral (14) becomes zero. In this case, the quantity $G^2(\mathbf{H},t)$, measured experimentally, is equal to integral (13) and is connected in this manner with the time average of the electron density. The term in Δ_T^2 is unimportant.

If (15) does not apply, then, to determine the average electron density, it is necessary to know how to separate $\mathfrak{F}^2(\mathbf{H})$ from $\mathfrak{F}^2_{T}(\mathbf{H})$.

We shall see later in which cases this separation can be carried out with assurance. In any case, such a separation is possible on the basis of experiment, that is, by studying x-ray scattering at different temperatures. It is clear that by lowering the temperature $\mathfrak{F}^2_{T}(\mathbf{H})$ becomes smaller compared to $\mathfrak{F}^2(\mathbf{H})$

In the majority of cases the correlation between $\Delta_T(\mathbf{R},t)$ and $\Delta_T(\mathbf{R}',t)$ is negligible; therefore $\mathfrak{F}^2_{T}(\mathbf{H}) \ll \mathfrak{F}^2(\mathbf{H})$ as a rule.

It follows from (14) that the study of $\mathfrak{F}^2_{T}(\mathbf{H})$ is interesting in its own right. One can hope to obtain information concerning the character of the thermal vibrations from a knowledge of this function. Up to now this has been done mainly for the simplest cubic crystals ([4] ch. III). For structure analysis, $\mathfrak{F}^2_{T}(\mathbf{H})$ is of interest only insofar as it is necessary to know how to eliminate this part of the scattering from (12).

It is the function $\mathfrak{F}^2(\mathbf{H})$ defined in (13), which determines the average electron density of the object, that is basically of interest in this book.

Let us call the amplitude of scattering of the average electron density the structure amplitude of the object

$$\mathfrak{F}(\mathbf{H}) = \int \bar{\rho}(\mathbf{R})\, e^{i2\pi\mathbf{HR}} dv. \tag{16}$$

Let us note that

$$\overline{G(\mathbf{H}, t)} = \mathfrak{F}(\mathbf{H}). \tag{17}$$

Then let us call the square of the modulus of the structure amplitude the structure factor. It is obvious that

$$\overline{G^2(\mathbf{H}, t)} \neq \mathfrak{F}^2(\mathbf{H}). \tag{18}$$

In fact

$$\mathfrak{F}^2(\mathbf{H}) = \mathfrak{F}(\mathbf{H})\,\mathfrak{F}^*(\mathbf{H}) = \iint \bar{\rho}(\mathbf{R})\,\bar{\rho}(\mathbf{R}')\, e^{i2\pi\mathbf{HR}} e^{-i2\pi\mathbf{HR}'} dv\, dv', \tag{19}$$

i.e., it is a part of the intensity $G^2(H,t)$ [see (12)]. The average, $\overline{G^2(H,t)}$ is, on the other hand equal to the full value of the intensity. Only when (15) is fulfilled is $\overline{G^2(H,t)} = \mathfrak{F}^2(H)$ Thus, the structure factor [but not $\overline{G^2(H,t)}$] characterizes the intensity of scattering of the average electron density.

Thus, if it has been possible to isolate experimentally the expression for $\mathfrak{F}^2(H)$ from (12), there arises the possibility of finding the time average of the electron density of the object. In this book, we shall examine the question of determining the time average of the electron density. For this reason (16) and (19) are basic for us.

Comparing (16) with (I, 11), we see that

$$\mathfrak{F}(\mathbf{H}) = \Phi(\bar{\rho}), \tag{20}$$

i.e., the structure amplitude is the Fourier transform of the time average electron density. Consequently,

$$\bar{\rho}(\mathbf{R}) = \Phi^{-1}(\mathfrak{F}), \tag{21}$$

i.e.,

$$\bar{\rho}(\mathbf{R}) = \int \mathfrak{F}(\mathbf{H})\, e^{-2\pi i \mathbf{HR}} d\tau. \tag{22}$$

Using (I, 52), we note that the structure factor (19) is the Fourier transform of the self-convolution of the average electron density.

$$\mathfrak{F}^2(\mathbf{H}) = \Phi S(\bar{\rho}', \bar{\rho}) \tag{23}$$

or

$$\int \bar{\rho}(\mathbf{R})\, \bar{\rho}(\mathbf{R}+\mathbf{R}')\, dv = \Phi^{-1}(\mathfrak{F}^2), \tag{24}$$

i.e.,

$$\int \bar{\rho}(\mathbf{R})\, \bar{\rho}(\mathbf{R}+\mathbf{R}')\, dv = \int \mathfrak{F}^2(\mathbf{H})\, e^{-2\pi i \mathbf{HR}} d\tau, \tag{25}$$

or, taking into account (I, 57),

$$S(\bar{\rho}', \bar{\rho}) = 2\int \mathfrak{F}^2(\mathbf{H}) \cos 2\pi \mathbf{HR}\, d\tau. \tag{26}$$

The relationships (22) and (26) connect the experimental data on the scattering of x-rays with the average electron density of the object.

We see that the problem of finding ρ (**R**) is not trivial, even when it is possible to separate the structure factor from the complete intensity. The right side of (26) can be found directly by experiment, but it allows one to compute for the object, not the electron density itself, but only the self-convolution of $\bar{\rho}$(**R**). Formula (22) makes it possible to find $\bar{\rho}$(**R**), but only the absolute magnitudes of the quantities $\mathfrak{F}$ (**H**) that occur in the right side of (22) are determined by experiment.

The tasks of the theory of x-ray structure analysis are the search for methods of determining the phases of $\mathfrak{F}$ (**H**) in order to use (22), and also for finding ways of computing $\bar{\rho}$ (**R**) from its self-convolution, known from (26).

The electron density $\bar{\rho}$ (**R**) of the object in the majority of cases determines uniquely the form of the self-convolution S($\bar{\rho}$, $\bar{\rho}$). Usually, there is no difficulty in choosing the correct structure from several homometric ones [this nomenclature has been given to structures which have different $\bar{\rho}$ (**R**), but which give identical self-convolutions]. In any case, the most substantial features of the structure are determined by the self-convolution of the electron density.

The intensity of scattering is the Fourier transform of the self-convolution of $\bar{\rho}$(**R**). The appearance of the intensity distribution $\mathfrak{F}^2$ (**H**) in reciprocal space allows the immediate computation of the self-convolution of the electron density as a function of the coordinates of physical space.

Even a qualitative examination of the function $\mathfrak{F}^2$ (**H**) allows one to come to important conclusions about the structure. We shall call the appearance of the distribution $\mathfrak{F}^2$ (**H**) the image of the function $\bar{\rho}$(**R**) in reciprocal space.

4. The Object Considered as a System of Atoms

Experimental results, as we have seen above, represent the average electron density of the object. Nevertheless, it is quite possible to consider the structure as an aggregate of atoms. The function $\bar{\rho}$(**R**) can always be represented in the form of a sum.

$$\bar{\rho}(\mathbf{R}) = \sum_j \bar{\rho}_j (\mathbf{R} - \mathbf{R}_j),$$

where $\bar{\rho}_j$ (**R**) is an atomic function, and $\mathbf{R}_j$ is the coordinate of the "center" of the atom. Naturally, such a representation cannot be unique. Therefore, the concept of the "center" of the atom needs to be defined more accurately.

The coordinates R_j may be found objectively only if the forms of the atomic functions of all the atoms are known.

Let us see what possibilities there are of computing the atomic functions from the experimental data.

The Fourier transform of the electron density can be written as

$$\mathfrak{F}(\mathbf{H})=\sum_{j}\int\bar{\rho}_j(\mathbf{R}-\mathbf{R}_j)\,e^{i2\pi\mathbf{HR}}dv,$$

i.e.,

$$\mathfrak{F}(\mathbf{H})=\sum_{j}f_j(\mathbf{H})\,e^{i2\pi\mathbf{HR}_j}. \tag{27}$$

The transform of the atomic function

$$f_j(\mathbf{H})=\int\bar{\rho}_j(\mathbf{R})\,e^{i2\pi\mathbf{HR}}dv \tag{28}$$

is called the atomic factor.

The intensity of scattering in this representation will have the form

$$\mathfrak{F}^2(\mathbf{H})=\sum_{j}\sum_{j'}f_j(\mathbf{H})\,f_{j'}(\mathbf{H})\,e^{i2\pi\mathbf{H}(\mathbf{R}_j-\mathbf{R}_{j'})}. \tag{29}$$

As is apparent from (28), the atomic functions could be computed by a Fourier transform, if there were a possibility of evaluating the atomic factor $f_j(\mathbf{H})$. This computation is impossible, since the form of $f_j(\mathbf{H})$ is affected not only by the kind of atom, but also by the structure of the object. The influence of the structure of the object is manifested first by the interatomic binding, and second, by the presence of thermal vibrations.

Thus, in principle $f_j(\mathbf{H})$ can be determined only after an experiment on the form of the electron density has been carried out. This is of no interest, since the goal of structure analysis is precisely the study of this electron density.

Since the form of $\bar{\rho}_j(\mathbf{R})$ is not known to us in advance, the determination of the "center" of the atom must be sought in the pattern of the electron density itself. In the future, by the centers of the atoms will be understood the coordinates of the maxima of $\bar{\rho}(\mathbf{R})$.

An important experimental fact is the following: Within the limits of experimental error any structure whatever can be represented as a sum of spherically symmetric functions. From this, it follows that spherically symmetric atomic factors $f_j(H)$ can be used for structure computations.

According to (I, 18) $R\bar{\rho}_j(R)$ and $Hf_j(H)$ are a pair of sine transforms.

The atomic function $\bar{\rho}_j(R)$ is indubitably a bell shaped curve. As a rule it is possible to give a good representation of an atomic function by (I, 21), or sometimes by (I, 20). Correspondingly, the atomic factors can be represented as

$$Ze^{-\alpha H^2} \quad \text{or} \quad Ze^{-\alpha |H|},$$

where Z is the atomic number of the atom (when H = 0, all the electrons scatter in phase and f = Z).[4]

It is sufficiently obvious that the values of α must differ for atoms with different numbers of electrons. The larger Z is, the steeper is the atomic function, and the greater the slope of the curve of the atomic factor. This follows from McWeeny's calculations (see below), and is confirmed by experiment.

If the atoms are close in atomic number, it is possible to assume that the atomic functions and factors will be similar. This enables us to use the relationship

$$f_j = Z_j \hat{f}, \tag{30}$$

where $\hat{f}$ is the unitary atomic factor.

To this approximation the equation for the amplitude of scattering takes the form

$$\mathfrak{F}(H) = \hat{f}(H) \sum Z_j e^{i2\pi \mathbf{H} \mathbf{R}_j}. \tag{31}$$

The $\mathfrak{F}$, computed from this formula can deviate by many tens of percent from the true ones. However, for different directions of H these deviations occur at random. Therefore, the mean squares (over different directions of H) of both sides of the equation agree well. And it is this that serves as the pri-

[4]Since we use these approximations outside of the region close to H = 0, it is sometimes convenient to multiply these expressions by a factor close to unity.

mary method for finding $\hat{f}$ from experiment (see chapter III, section 8).

In exceptional cases, during the last stages of a structure analysis, the computation of atomic factors can be of some (very limited) interest.

Let us denote by f_0 the scattering amplitude of an atom at its equilibrium position. Let the thermal displacement of the atom be Δ. Then the scattering amplitude will be $\overline{f_0 e^{i2\pi \mathbf{H}\Delta}}$.

We shall be interested in this time average of the scattering amplitude, equal to $\overline{f_0 e^{i2\pi \mathbf{H}\Delta}}$.

Let us denote the projection of the vector Δ on the vector H by z.

It is this projection of Δ that determines the scattering at a given angle. Then the instantaneous amplitude of scattering is written

$$f_0 e^{i2\pi Hz}.$$

Let the probability of a displacement of the atom by magnitude z be w(z). Let us assume a Gaussian distribution of the displacements of the atom, i.e.,

$$w(z) = \frac{1}{\sqrt{2\pi D}} e^{-\frac{z^2}{2D}},$$

where the dispersion D of the distribution means the mean square deviation $u_{\mathbf{H}}^2$ in the direction of the reciprocal vector.

And so, the instantaneous amplitude is

$$f_0 e^{i2\pi Hz},$$

and the average amplitude of interest to us is equal to

$$f_0 \int_{-\infty}^{\infty} w(z)\, e^{i2\pi Hz} dz = f_0 e^{-M},$$

where

$$M = 8\pi^2 u_{\mathbf{H}}^2 \left(\frac{\sin \vartheta}{\lambda}\right)^2 = 2\pi^2 (\Delta \mathbf{H})^2. \tag{32}$$

The last expression can be also written as

$$M = 2\pi^2 (\Delta_x H_x + \Delta_y H_y + \Delta_z H_z)^2,$$

where Δ_x, Δ_y, and Δ_z are constants which have the meaning of vibration projections on the axes of the reciprocal lattice. In order to simplify the analysis we have confined ourselves to the expression for a crystal of orthorhombic symmetry.

It is not difficult to deduce that in a general case the expression for M can be brought into the form

$$M = (h^2\alpha + k^2\beta + l^2\gamma + 2hk\delta + 2kl\varepsilon + 2hl\eta), \tag{33}$$

thus characterizing the vibrational anisotropy by six constants.

If there is a basis for considering the ellipsoid of vibration to be uniaxial, then M may be represented in the form

$$M = a \cos^2 \varphi + b \sin^2 \varphi, \tag{34}$$

where φ is measured from the axis of the ellipsoid, which, generally speaking, can be found experimentally.

There is no way to compute these constants theoretically. The construction of a theory is possible for simple cubic crystals ([5] ch. V) only.

The question of the atomic factor f_0 for atoms at rest was recently given a new analysis by McWeeny. In his first paper [6] this author computed anew the atomic factors for light atoms. For atoms that have no spherical symmetry (B, C, O, F), McWeeny calculated the values of the atomic factor $f_0^{\parallel}$ and $f_0^{\perp}$ for scattering in the directions parallel and perpendicular to the axis of symmetry of the electron cloud of these atoms. If the vector **H** forms an angle φ with the direction of the axis of symmetry, then the atomic factor should be computed according to the formula

$$f_0 = f^{\parallel} \cos^2 \varphi + f^{\perp} \sin^2 \varphi. \tag{35}$$

As an approximation one may use the formula

$$\bar{f}_0 = \frac{1}{3} f^{\parallel} + \frac{2}{3} f^{\perp}. \tag{36}$$

TABLE 1. Atomic Scattering Factor of Light Atoms.

Element	$\sin\vartheta/\lambda$*	0	0.025	0.05	0.075	0.1	0.125
H	f	1.0	0.952	0.829	0.664	0.514	0.383
He	f	2.0	1.964	1.869	1.722	1.544	1.353
Li	f	3.0	2.709	2.263	1.950	1.786	1.675
Be	f	4.0	3.742	3.150	2.553	2.129	1.878
B	$f^{\parallel}$	5.0	4.710	3.982	3.130	2.423	1.941
	$f^{\perp}$	5.0	4.802	4.297	3.647	3.042	2.559
C	$f^{\parallel}$	6.0	5.846	5.417	4.874	4.172	3.565
	$f^{\perp}$	6.0	5.786	5.213	4.492	3.639	2.941
C (covalent bonds)	f	6.0	5.803	5.271	4.581	3.803	3.137
N	f	7.0	6.822	6.374	5.703	4.890	4.164
O	$f^{\parallel}$	8.0	7.850	7.396	6.643	5.791	4.915
	$f^{\perp}$	8.0	7.868	7.526	6.865	6.144	5.383
F	$f^{\parallel}$	9.0	8.901	8.540	8.024	7.359	6.630
	$f^{\perp}$	9.0	8.889	8.457	7.834	7.072	6.235
Ne	f	10.0	9.928	9.540	8.991	8.301	7.520

*Sin ϑ/λ is expressed here in reciprocal atomic units x = 0.5282A (and not,

A modern table of atomic factors is given above (Table 1).

In his second paper [7] McWeeny has shown that for bonded atoms it is possible to introduce the concept of an effective atomic factor

$$f_{0e} = f_0 + f_{0AB}, \tag{37}$$

where f_{0AB} is the atomic factor of the binding electrons, the density of which can be assumed to be smeared over the space which surrounds the bonded atoms A and B. It is clear that f_{0AB} is a function of the direction of the reciprocal vector. McWeeny showed that the effect of the binding electrons is a maximum in case the vector **H** coincides with the bond direction. When these directions are perpendicular the effect of f_{0AB} is negligible. It should be noted that the maximum deviation from f_0 and the effect of orientation are present for the average scattering angles.

In the third paper of this series [8], the author has devised a way for computing f_{0AB} for concrete cases, and for evaluating the accuracy of these calculations, which was found to be 5%.

0.15	0.175	0.2	0.25	0.3	0.4	0.5	0.6	0.7	0.8
0.281	0.205	0.150	0.087	0.048	0.019	0.009	0.004	0.002	0.001
1.164	0.987	0.829	0.576	0.397	0.194	0.101	0.056	0.033	0.021
1.572	1.462	1.348	1.119	0.908	0.576	0.362	0.230	0.150	0.100
1.736	1.644	1.569	1.424	1.266	0.955	0.695	0.494	0.352	0.253
1.675	1.539	1.480	1.435	1.383	1.195	0.966	0.759	0.586	0.449
2.214	1.981	1.824	1.626	1.486	1.227	0.977	0.764	0.588	0.450
3.051	2.646	2.340	1.945	1.717	1.435	1.208	0.975	0.814	0.658
2.421	2.052	1.813	1.576	1.398	1.342	1.170	0.959	0.807	0.653
2.629	2.255	1.999	1.711	1.566	1.368	1.169	0.973	0.795	0.654
3.489	2.892	2.507	2.043	1.697	1.464	1.316	1.156	0.994	0.841
4.092	3.375	2.790	2.002	1.614	1.365	1.300	1.221	1.110	0.988
4.649	3.987	3.422	2.595	2.102	1.647	1.448	1.297	1.150	1.010
5.885	5.164	4.590	3.454	2.718	1.941	1.615	1.431	1.283	1.146
5.393	4.610	3.979	2.826	2.145	1.555	1.388	1.304	1.212	1.105
6.699	5.888	5.128	3.839	2.951	1.906	1.549	1.386	1.303	1.207

as is usual, in reciprocal A).

It is regrettable that McWeeny's careful work is almost entirely of academic interest. There are two reasons for this: 1) The majority of electrons take part in the formation of a spherically symmetric cloud. 2) The anisotropy of thermal motion has a more pronounced effect than has the anisotropy of the electron arrangement near the direction of a chemical bond.

5. The Object as a System of Particles

The computation of the scattering intensity from a given structure is always possible, but, unfortunately, only rarely of practical interest. This concerns primarily the objects most difficult for structure analysis, namely, systems of particles that do not form a regular lattice.

Let $\rho_k(\mathbf{R})$ be the electron density of the kth particle. Its amplitude of scattering is

$$\mathfrak{F}_k(\mathbf{H}) = \Phi(\rho_k). \tag{38}$$

The amplitude of scattering by a system of particles is, in the most general case, equal to

$$\mathfrak{F}(\mathbf{H}) = \sum_{k=1}^{N} \mathfrak{F}_k(\mathbf{H})\, e^{2\pi i \mathbf{H}\mathbf{R}_k}, \tag{39}$$

where $\mathbf{R}_k$ gives the coordinates of the "centers" of the particles; the N particles in general differ from one another in structure and orientation.

The scattering intensity of the whole system has the form

$$\mathfrak{F}^2(\mathbf{H}) = \sum_{k=1}^{N} \mathfrak{F}_k(\mathbf{H})\, \mathfrak{F}_k^*(\mathbf{H}) + \sum\sum_{j \neq k} \mathfrak{F}_j(\mathbf{H}) \mathfrak{F}_k^*(\mathbf{H})\, e^{2\pi i \mathbf{H}(\mathbf{R}_j - \mathbf{R}_k)}. \tag{40}$$

The double sum in this expression may be called a correlative term. It is determined by the character of the relative arrangement of the particles.

If a system of particles is given between which correlation can be neglected, then the scattering intensity will be determined by the expression

$$\sum_k \mathfrak{F}_k(\mathbf{H})\, \mathfrak{F}_k^*(\mathbf{H}). \tag{41}$$

If there is no correlation, it is always possible to consider the total intensity as the sum of the intensities scattered by particles of a single kind. Let us assume, therefore, that $\mathfrak{F}_k(\mathbf{H})$ are the amplitudes of scattering of the same, but differently oriented, particles.

Let us denote by $P_k(\mathbf{R})$ the self-convolution of the electron density of one particle. The scattering intensity of one particle is equal to

$$\mathfrak{F}_k(\mathbf{H})\, \mathfrak{F}_k^*(\mathbf{H}) = \Phi(P_k), \tag{42}$$

and the total intensity for the system of particles is

$$\sum \Phi(P_k). \tag{43}$$

When there is no correlation, the distribution of the "centers" of the particles does not affect the intensity. In this case the intensity scattered by the accumulation of differently oriented particles is equal to the superposition of the intensities scattered by one particle which occupies successively all possible orientations in space.

The orientation of the particle in space is characterized in the general case by three angles φ, Ψ, x, which can be assigned in direct as well as in reciprocal space. Let dN (φ, x, Ψ) be the number of particles with an orientation within the limits from φ to $\varphi + d\varphi$, from x to x + dx and from Ψ to $\Psi + d\Psi$. The quantity dN can be considered as the time a particle remains in the corresponding orientation.

To the particle corresponds its own image in reciprocal space. A rotation of the particle produces a corresponding rotation of its image. Thus, finding the intensity scattered by the system of particles is reduced to the integration of the image of one particle:

$$\mathfrak{F}^2(\mathbf{H}) = \int \Phi(P)\, dN. \tag{44}$$

If the angular distribution of the particles is assigned, then the determination of $\mathfrak{F}^2(\mathbf{H})$ presents a computational problem.

Problems in which all $\mathbf{R}_k$ and $\mathfrak{F}_k$ are assigned are of no practical importance. Correlation is either neglected or taken into account by characterizing the system with numerical parameters or distribution functions which give the probability of a definite short range order.

To take correlation into account means to find the dependence of the intensity of the scattered radiation on such parameters or functions. Correlation can be taken into account in practice only for the one-dimensional case, and for spherical particles when there is a dependence on $|\mathbf{R}|$ only.

The computational methods for such problems are discussed in the author's book "X-ray Structural Analysis of Microcrystalline and Amorphous Substances" (see chapter II).

6. The Form Factor

As will be seen below, it is very useful to separate out the effect of the external form of the object on the x-ray scattering. Let us assume that the distribution of electron density of the object is given by the function $\rho(R)$ which has values different from zero in unlimited space. If $\varphi(R)$ is a function satisfying the condition $\varphi(\mathbf{R}) = 0$ outside the object, and $\varphi(\mathbf{R}) = 1$ inside the object, then the electron density of the body is given by the product $\varphi(\mathbf{R})\rho(\mathbf{R})$.

According to (20) the amplitude of scattering of a finite object is $\Phi(\varphi\bar{\rho})$. On the other hand, according to (I, 35) this Fourier transform is equal to the convolutions of the transforms of $\bar{\rho}(\mathbf{R})$ and $\varphi(\mathbf{R})$ Denoting as usual $\Phi(\bar{\rho}) = \mathfrak{F}$ and introducing the notation

$$s(\mathbf{H}) = \Phi(\varphi) = \int \varphi(\mathbf{R})\, e^{i2\pi \mathbf{HR}}\, dv, \tag{45}$$

we obtain for the amplitude of scattering of a finite object the expression

$$\mathfrak{F}_{\text{fin}}(\mathbf{H}) = \int s(\mathbf{H} - \mathbf{H}')\, \mathfrak{F}(\mathbf{H}')\, d\tau'. \tag{46}$$

The function $s(\mathbf{H})$ is called the form factor.

As follows from the definition, the form factor is the amplitude of scattering by an object with a uniform electron density equal to unity; $s(\mathbf{H})$ has the dimensions of volume.

To normalize the form factor it is useful to note the relationship

$$\int |s|^2 d\tau = \int |\varphi|^2 dv = VN, \tag{47}$$

which follows from (I, 15). Here N is the volume in relative units, and V is the volume of the metric.

The scattering intensity of a finite object must be calculated by multiplying $\mathfrak{F}_{\text{fin}}(\mathbf{H})$ by its complex conjugate.

The effect of the form factor on reciprocal space is especially evident in the case of a particle with a constant electron density.

The amplitude of scattering is equal to

$$\mathfrak{F}(\mathbf{H}) = \bar{\rho} s(\mathbf{H}),$$

the intensity is

$$\mathfrak{F}^2(H) = \bar{\rho}^2 s^2(H). \tag{48}$$

These formulas have a general meaning if $\bar{\rho}$ is taken as the average value of the electron density not only over time but also over volume.

On the basis of (47) we obtain for the integrated intensity

$$I = \bar{\rho}^2 \int s^2(\mathbf{H})\, d\tau = \bar{\rho}^2 V N. \tag{49}$$

Let us examine some special cases of the form factor.

<u>The Spherical Particle.</u> In this case

$$s(H) = \int_0^R \int_0^\pi \int_0^{2\pi} e^{i2\pi HR\cos\alpha} R^2 \sin\alpha \, dR\, d\alpha\, d\varphi.$$

Performing the integration over α and φ we obtain

$$s(H) = \frac{2}{H} \int_0^R R \sin 2\pi HR dR = \frac{4}{3}\pi R^3 \Phi(u). \tag{50}$$

where

$$\Phi(u) = 3\frac{\sin u - u\cos u}{u^3},$$

and

$$u = 2\pi HR. \tag{51}$$

The appearance of the square of this function is shown in Fig. 3. To a sufficient approximation it is possible to consider that $\Phi^2(u)$ represents a bell-shaped curve. The secondary diffraction maxima decrease rapidly. It is interesting to note that in spite of the negligible part of the intensity that goes into the secondary maxima, it is still possible to observe them experimentally by studying the scattering at small angles from the initial beam.

As follows from the curve with $u \approx 3$, the intensity of the scattered radiation, proportional to $\Phi^2(u)$, is about 100 times smaller than the intensity in the maximum.

Consequently, the image in intensity space of a spherical particle of radius R will be a sphere with the effective radius

$$H \approx \frac{3}{2\pi R} \approx \frac{1}{2R} \tag{52}$$

If $R > 100$ A then the radius of the sphere in reciprocal space will be

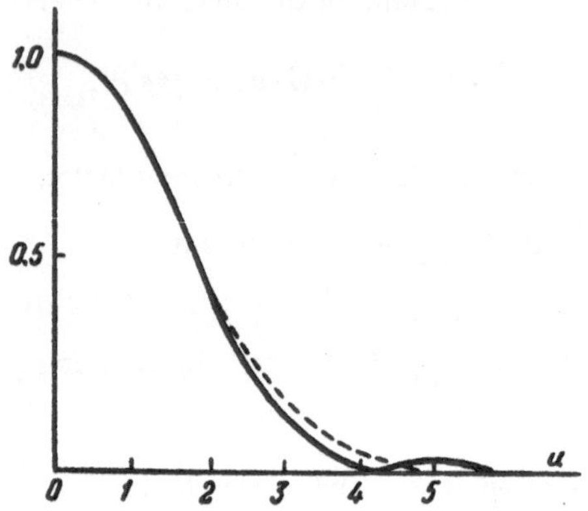

Fig. 3. Curve of $\Phi^2(u)$ (solid line) and of the function e^{-bu^2} closest to it (dotted).

smaller than that indicated above. Therefore, under usual conditions of exposure, the intensity image of an object which has a radius greater than 100 A, is a point. With special exposure methods this limit can be increased by one order of magnitude.

The Spheroidal Particle. Let us now solve a more general problem, namely, the calculation of the intensity of scattering from a spheroidal particle (ellipsoid of revolution) with an axial ratio equal to w. Let Oz be the axis of symmetry perpendicular to the circular section. Let us denote by ϵ the angle formed by the vector **H** with the Oz axis. The amplitude of scattering, now dependent on the relative orientation of the initial beam and the ellipsoid, is equal to

$$A = \bar{\rho} \iiint e^{i2\pi HX} dv,$$

where X is the projection of the vector **r**, linking the origin of coordinates with an arbitrary point of the ellipsoid in the direction **H**.

To carry out the integration, let us break up the ellipsoid into infinitely thin layers perpendicular to **H**. Let us draw about the center of the ellipsoid an auxiliary sphere of radius a, the circular section of which coincides with the circular section of the ellipsoid.

One may substitute a point of the circle for each point of the ellipsoid by dividing the ordinates of the ellipsoidal body by w. To the points of the elliptical shape of the ellipsoidal body NM, corresponds the spherical shell

N'M' of the sphere (Fig. 4). If the angle of the normal to the elliptical layer with the axis Oz is ϵ, and the angle of the normal to the spherical layer is α, then

$$\operatorname{tg}\varepsilon = w \operatorname{tg}\alpha.$$

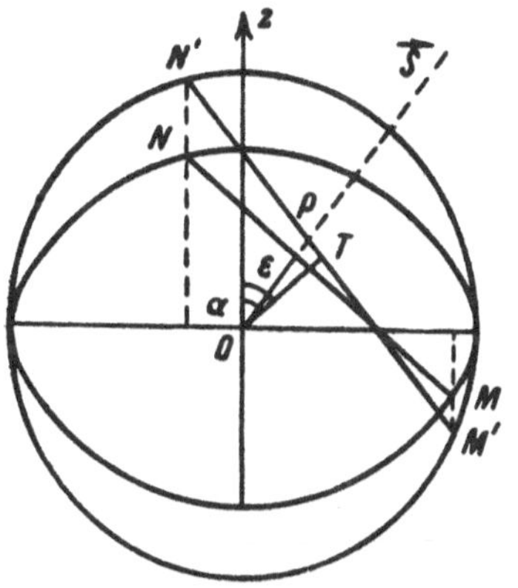

Fig. 4. Concerning the calculation of the intensity of scattering from an ellipsoidal particle. OP is the normal to the plane of the ellipsoidal layer, OT is the normal to the plane of the spherical layer.

Let us denote by t = OT the distance along the normal to the spherical layer. The radius of the circumference of this layer is equal to $\sqrt{a^2 - t^2}$. The volume of the spherical layer is equal to $\pi (a^2 - t^2)dt$. The volume of the elliptical layer that is of interest to us will differ from the volume of the spherical layer by a factor w

$$dv = \pi w (a^2 - t^2)\, dt.$$

Through the same parameter t, we can express the quantity X = OP — the length of the normal dropped from the origin of coordinates onto the elliptical section

$$X = \frac{wt}{\sqrt{\cos^2\alpha + w^2 \sin^2\alpha}}.$$

The desired expression for the amplitude acquires the form

$$A=\int_{-a}^{+a}\pi\left(a^2-t^2\right)we^{-i2\pi H\frac{wt}{\sqrt{\cos^2\alpha+w^2\sin^2\alpha}}}dt. \tag{53}$$

The integral is easily taken and reduced to the form $A = N\Phi(u)$, or, for the intensity, we have again $I = N^2\Phi^2(u)$, where $\Phi(u)$ has its former value (51), but

$$u=\frac{2\pi Hwa}{\sqrt{\cos^2\alpha+w^2\sin^2\alpha}}=2\pi Ha\sqrt{\sin^2\varepsilon+w^2\cos\varepsilon}.$$

We have obtained an expression for the intensity of the scattered radiation for the orientation of the ellipsoid in which the vector H forms an angle ϵ with the principal axis.

Let us consider again the value $u = 3$ as the practical limit of the intensity image in reciprocal space of a particle. Then the equation of the boundary surface of the region of reciprocal space will have the form

$$3=2\pi aH\sqrt{\sin^2\varepsilon+w^2\cos^2\varepsilon}. \tag{54}$$

This is also the equation of an ellipsoid of rotation. The intensity image of an ellipsoidal particle is an ellipsoid. Since

$$H=\frac{3}{2\pi a\sqrt{1+\cos^2\varepsilon\left(w^2-1\right)}}, \tag{55}$$

then, evidently, the length of the rotation axis of the ellipsoid ($\epsilon = 0$) is equal to $\frac{3}{2\pi aw}$. If the particle is a lengthened ellipsoid of rotation ($w > 1$) then its image in reciprocal space is a flattened ellipsoid and vice versa. The lengths of the axes of the ellipsoid image are inversely proportional to the lengths of the axes of the particles.

The extreme cases of the last formula are the disk and the rod. If the particle has the shape of an infinitely thin disk of radius $\underline{a}$ ($w = 0$), then the intensity image is an infinitely long rod. In fact, when the angle $\epsilon \to 90°$ the radius of the image $H \to \infty$. The radius of the rod will be as before $\frac{3}{2\pi a}$. If the particle has the form of an infinitely thin rod of length $L = aw$, then, assuming $a \to 0$ and $w \to \infty$, we obtain $H = 0$ for all ϵ, except $\epsilon = 0°$. In the same case $H = \frac{3}{2\pi L}$, i.e., in agreement with the reversibility of physical and reciprocal space, the image of such a rod is an infinite disk of thickness inversely proportional to the length of the rod-shaped particle.

The Parallelepiped. If N_1, N_2, and N_3 are the relative lengths of the edges of a parallelepiped, then the form factor can be written down as

$$s(\mathbf{H}) = V \int_{-\frac{1}{2}N_1}^{+\frac{1}{2}N_1} \int_{-\frac{1}{2}N_2}^{+\frac{1}{2}N_2} \int_{-\frac{1}{2}N_3}^{+\frac{1}{2}N_3} e^{i2\pi(x\xi+y\eta+z\zeta)}\, dx\, dy\, dz.$$

Let us recall that x,y,z are relative (dimensionless) coordinates, that V is the volume of the metric, and

$$\int_{-\frac{1}{2}N_1}^{\frac{1}{2}N_1} e^{i2\pi x\xi} dx = \frac{\sin \pi\xi N_1}{\pi\xi}.$$

Integration gives

$$s(\mathbf{H}) = VN \frac{\sin \pi\xi N_1}{\pi\xi N_1} \frac{\sin \pi\eta N_2}{\pi\eta N_2} \frac{\sin \pi\zeta N_3}{\pi\zeta N_3}, \qquad (56)$$

where $VN = VN_1N_2N_3$ is the volume of the object. The scattering intensity is proportional to $s^2(\mathbf{H})$. The shape of the function $\frac{\sin^2 t}{t^2}$ is well known. Its secondary maxima are very small in comparison with the principal maximum. The first zero occurs when $\xi = \frac{1}{N_1}$, i.e., when $H_X = \frac{1}{A}$, where A is the absolute length of the edge.

Thus, a parallelepiped of dimensions 1/A, 1/B, 1/C can be taken as the image of a parallelepiped of dimensions A, B, C. It is obvious that this is not exact;[5] however, the secondary maxima here also are too small and are unimportant.

7. Infinite δ Lattices

The lattice is a descriptive name for an ideal figure in which the density of the substance has periodicity.

[5]A detailed examination of the question pertaining to the form of the intensity image for different polyhedra is given in James' book [5] and, also A. L. Patterson's paper [9].

Let us call abstract figures which represent infinite periodic sequences of points, lines, or planes δ lattices. The distribution of a "substance" in δ lattices will be described by "lattice-like" functions $\mathfrak{G}(\mathbf{R})$.

The following one-dimensionally periodic lattices are possible: 1) a series of points, 2) a series of lines, and 3) a series of planes. The following lattice-like functions will correspond to them

$$1)\ \mathfrak{G}(\mathbf{R}) = \frac{Z}{V}\delta(x)\delta(y)\sum_{k=-\infty}^{+\infty}\delta(z - z_k);$$

$$2)\ \mathfrak{G}(\mathbf{R}) = \frac{Z}{V}\delta(y)\sum_{k=-\infty}^{+\infty}\delta(z - z_k);$$

$$3)\ \mathfrak{G}(\mathbf{R}) = \frac{Z}{V}\sum_{k=-\infty}^{+\infty}\delta(z - z_k),$$

where $z_{k+1} - z_k = 1$.

The formulas are written for a series of points distributed along the z axis, for a series of lines lying in the xz plane, and for a series of planes perpendicular to the z axis; $\delta(x)$ is the notation for the δ function, equal to zero for all values of x except for $x = 0$.

To describe the lattice it is natural to choose the metric of space in such a way that the periods equal the axial unit lengths.

To each "node" of the δ lattice "belongs" a definite part of space (an infinite layer, an infinite rod, and so forth). Denoting the volume of this part of space by NV, where V is the volume of the metric, and the number of electrons for this volume by NZ (Z is the number of electrons in the volume V), we can formally assign to the δ lattices a "weight" equal to Z/V, i.e., equal to the electron "density." Attributing such "weight" to the node of the lattice is convenient for what comes later (see below).

Let us find the images of the distributions of the "substance" in reciprocal space. Let us call the images of the lattice functions Laue functions, L(H). By definition the concept of the image $L(\mathbf{H}) = \mathfrak{F}^2$, where $\mathfrak{F}$ is the scattering amplitude of an ideal δ lattice and is equal to the Fourier transform of $\mathfrak{G}(\mathbf{R})$.

For a series of points along z we obtain

$$\mathfrak{F}(\mathbf{H}) = \frac{Z}{V}\int e^{i2\pi\mathbf{HR}}\,\delta(x)\,\delta(y)\sum\delta(z - z_k)\,dv =$$
$$= Z\int e^{i2\pi(\xi x+\eta y+z\zeta)}\,\delta(x)\,\delta(y)\sum\delta(z - z_k)\,dx\,dy\,dz.$$

Integration along x and y gives unity. Consequently,

$$\mathfrak{F}(\mathbf{H}) = Z\sum_k e^{2\pi i\zeta z_k}.$$

From this

$$L(\mathbf{H}) = Z^2\sum_k\sum_{k'} e^{2\pi i\zeta(z_k - z_{k'})}.$$

The double sum is a geometric progression, the computation of which again reduces to a sum of δ functions. In fact, this sum is equal to zero for all values of ζ, except for those where ζ is equal to a whole number. Therefore,

$$L(\mathbf{H}) = Z^2\sum_{l=-\infty}^{+\infty}\delta(\zeta - \zeta_l).$$

Consequently, the image of a series of points arranged along the axis of object space is a series of planes arranged parallel to the $\xi\eta$ plane of reciprocal space with a period equal to unity. Let us recall that the metrics of reciprocal space are connected with those of physical space by the conditions (I, 8). The weight of the Laue function, proportional to the square of the number of electrons, has the following meaning: it is the scattering intensity for the direction **H**.

For the second one-dimensional lattice– a series of lines– we obtain

$$\mathfrak{F}(\mathbf{H}) = Z\int e^{i2\pi(\xi x+\eta y+\zeta z)}\,\delta(y)\,\Sigma\delta(z - z_k)\,dx\,dy\,dz.$$

Integration along y gives unity. Integration along x gives $\delta(\xi)$.[6] Thus

$$L(\mathbf{H}) = Z^2\delta^2(\xi)\sum_{-\infty}^{+\infty}\delta(\zeta - \zeta_l).$$

[6] $\int_{-\infty}^{+\infty} e^{2\pi i\xi x}\,dx$ is the classical example of the δ function.

We see that the image of a series of parallel lines lying in the xz plane and parallel to the x axis is a series of lines lying in the $\eta\zeta$ plane and parallel to the η axis.

Finally, the last example – a series of planes – gives

$$\mathfrak{F}(\mathrm{H}) = Z\int e^{i2\pi(\xi x+\eta y+\zeta z)}\,\Sigma\delta(z-z_k)\,dx\,dy\,dz =$$

$$= Z\delta(\xi)\,\delta(\eta)\,\Sigma e^{2\pi i\zeta z_k}.$$

From this

$$L(\mathrm{H}) = Z^2\delta^2(\xi)\,\delta^2(\eta)\sum_{l=-\infty}^{+\infty}(\zeta-\zeta_l),$$

i.e., the image is a series of points arranged along ζ.

In all the three cases of one-dimensional periodicity with a period of unity, there is a corresponding one-dimensional periodicity with a period of unity in reciprocal space. In all cases the images of the object and of the reciprocal space are orthogonal, that is, to special directions of one space correspond special planes of another and vice versa.

Let us examine now the two-dimensionally-periodic lattices.

A two-dimensional point system in the xy plane, based on a pair of axial vectors with the metrics a_1 and a_2, is described by the lattice function

$$\mathfrak{G}(\mathrm{R}) = \frac{Z}{V}\,\delta(z)\sum\delta(x-x_k)\sum\delta(y-y_k),$$

where

$$x_{k+1}-x_k=1 \quad\text{and}\quad y_{k+1}-y_k=1.$$

Carrying out the computation of the Laue function in the same manner as in the one-dimensional case, we obtain

$$L(\mathrm{H}) = Z^2\sum_h\delta(\xi-\xi_h)\sum_k\delta(\eta-\eta_k).$$

The image of a two-dimensional point lattice is a two-dimensional lattice of rods parallel to the ζ axis.

Evidently, the reverse is also true, i.e., the image of a rod lattice is a two-dimensional point lattice.

We emphasize that, as before, all the special planes are reflected in reciprocal space by straight lines and vice versa. The rod axes in one space correspond to the normals to two-dimensional lattices in the other space, etc.

The method just indicated makes it possible to construct easily the images of the other conceivable two-dimensional lattices also.

Let us consider the three-dimensional lattice. The lattice function can be represented as

$$\mathfrak{G}(\mathbf{R}) = \frac{Z}{V} \sum_m \delta(x - x_m) \sum_n \delta(y - y_n) \sum_p \delta(z - z_p),$$

or better, in vector form

$$\mathfrak{G}(\mathbf{R}) = \frac{Z}{V} \sum_L \delta(\mathbf{R} - \mathbf{r}_L), \tag{57}$$

where $\mathbf{r}_L = m\mathbf{a}_1 + n\mathbf{a}_2 + p\mathbf{a}_3$ ($\mathbf{a}_i$ are the axial unit vectors).

The scattering amplitude of the three dimensional lattice is

$$\mathfrak{F}(\mathbf{H}) = Z \int e^{2\pi i \mathbf{H}\mathbf{R}} \sum_L \delta(\mathbf{R} - \mathbf{r}_L)\, dx\, dy\, dz,$$

i.e.,

$$\mathfrak{F}(\mathbf{H}) = Z \sum_L e^{2\pi i \mathbf{H}\mathbf{r}_L} \tag{58}$$

and consequently,

$$L(\mathbf{H}) = Z^2 \sum_L \sum_{L'} e^{2\pi i \mathbf{H}(\mathbf{r}_L - \mathbf{r}_{L'})}, \tag{59}$$

but the difference of lattice vectors in (59) is also a lattice vector. According to (I, 7),

$$\mathbf{H}\mathbf{r}_{mnp} = m\xi + n\eta + p\zeta.$$

But the sum in (59) is equal to zero *in all cases*, except in those cases where the equations

$$m\xi = n_1, \quad n\eta = n_2, \quad p\zeta = n_3, \tag{60}$$

are fulfilled simultaneously, where n_1, n_2, and n_3 are whole numbers. In the latter cases the sum is equal to unity. Consequently, $L(\mathbf{H})$ differs from zero only in those, and in all those, points of reciprocal space, where ξ, η and ζ are whole numbers. In other words, the reciprocal space is a point lattice, the periods of which are the axial vectors of reciprocal space determined by the conditions (I, 7). This lattice is called the reciprocal lattice of the ideal crystal. Thus

$$L(\mathbf{H}) = Z^2 \Sigma \delta (\mathbf{H} - \mathbf{H}_{hkl}), \tag{61}$$

where

$$\mathbf{H}_{hkl} = h\mathbf{b}_1 + k\mathbf{b}_2 + l\mathbf{b}_3,$$

and hkl are the whole numbers. The factor Z^2 – the "weight" of the Laue function – means the scattering intensity for the $\mathbf{H}_{hkl}$ direction. In agreement with the general formula (23), the Laue function $L(\mathbf{H})$ is the Fourier transform of the self-convolution of the lattice function.

The self-convolution of $\mathfrak{G}(\mathbf{R})$ has the form

$$S(\mathfrak{G}', \mathfrak{G}) = \frac{Z^2}{V^2} \int \sum_{L} \delta(\mathbf{R}' - \mathbf{r}_L) \sum_{L'} \delta(\mathbf{R} + \mathbf{R}' - \mathbf{r}_{L'})\, dv'.$$

The integral (function of $\mathbf{R}$) has values different from zero only when $\mathbf{R} = \mathbf{r}_{L'} - \mathbf{r}_L$, i.e.,

$$S(\mathfrak{G}', \mathfrak{G}) = \frac{Z^2}{V} \sum_{L} \sum_{L'} \delta[\mathbf{R} - (\mathbf{r}_{L'} - \mathbf{r}_L)]. \tag{62}$$

If $\mathfrak{G}(\mathbf{R})$ is centrosymmetric [$\mathfrak{G}(\mathbf{R}) = \mathfrak{G}(-\mathbf{R})$; this means that the origin of the lattice is chosen at zero], then $S(\mathfrak{G}, \mathfrak{G}) = Z\mathfrak{G}$, i.e., the lattice function coincides in this case with its own convolution. The "weight" of a node of the convolution is Z times greater than the "weight" of a lattice node.

The Fourier transform of the convolution is equal to

$$\Phi[S(\mathfrak{G}, \mathfrak{G})] = \frac{Z^2}{V} \int \sum_{L} \sum_{L'} \delta[\mathbf{R} - (\mathbf{r}_{L'} - \mathbf{r}_L)]\, e^{2\pi i \mathbf{H}\mathbf{R}} dv.$$

The expression under the integral differs from zero only when $\mathbf{R} = \mathbf{r}_{L'} - \mathbf{r}_L$. The integral transforms into a double sum

$$Z^2 \sum_{L} \sum_{L'} e^{2\pi i \mathbf{H}(\mathbf{r}_{L'} - \mathbf{r}_L)}.$$

Comparing with (59), we see that, in fact,

$$\mathfrak{F}^2 = \Phi[S(\mathfrak{G}, \mathfrak{G})].$$

In conclusion, let us formulate a few theorems concerning any lattice functions. The proof is evident from the preceding.

1. The image in reciprocal space of the lattice function

$$\mathfrak{G}=\frac{Z}{V}\sum\delta(\mathbf{R}-\mathbf{r}_{mnp})$$

is also a lattice function

$$Z^2\,\Sigma\delta(\mathbf{H}-\mathbf{H}_{hkl}). \tag{63}$$

2. The convolution of an arbitrary function $\varphi(\mathbf{R})$ with a lattice function consists in duplicating $\varphi(\mathbf{R})$ at all the lattice nodes

$$S(\varphi,\mathfrak{G})=\frac{Z}{V}\int\varphi(\mathbf{R}')\sum\delta(\mathbf{R}-\mathbf{r}_{mnp}-\mathbf{R}')\,dv',$$

which gives

$$S(\varphi,\mathfrak{G})=Z\Sigma\varphi(\mathbf{R}-\mathbf{r}_{mnp}). \tag{64}$$

3. The self-convolution of a lattice function

$$S(\mathfrak{G},\mathfrak{G})=Z\cdot\mathfrak{G} \tag{65}$$

is also a lattice function.

8. Finite δ Lattices

We shall call the product of a lattice function $\mathfrak{G}(\mathbf{R})$ by a function $\varphi(\mathbf{R})$ (see section 7) a finite δ-space lattice when $\varphi(\mathbf{R})$ possesses the following properties: $\varphi(\mathbf{R}) = 1$ inside some region and $\varphi(\mathbf{R}) = 0$ outside of this region.

The scattering amplitude of such a figure is equal to

$$\mathfrak{F}(\mathbf{H})=\Phi(\mathfrak{G}\cdot\varphi).$$

But the Fourier transform of the product is the convolution of the corresponding transforms. Introducing, in the same way as on page 36, the form factor of the region occupied by the lattice $s = \Phi(\varphi)$, and remembering that the reverse Fourier transform of a lattice function is equal to

$$Z\Sigma\delta(\mathbf{H}-\mathbf{H}_{hkl}),$$

we obtain for the scattering amplitude

$$\mathfrak{F}=S\,[sZ\Sigma\delta(\mathbf{H}-\mathbf{H}_{hkl})],$$

i.e.,

$$\mathfrak{F}=Z\int s(\mathbf{H}-\mathbf{H}')\sum_{k}\delta(\mathbf{H}'-\mathbf{H}_{hkl})\,d\tau', \tag{66}$$

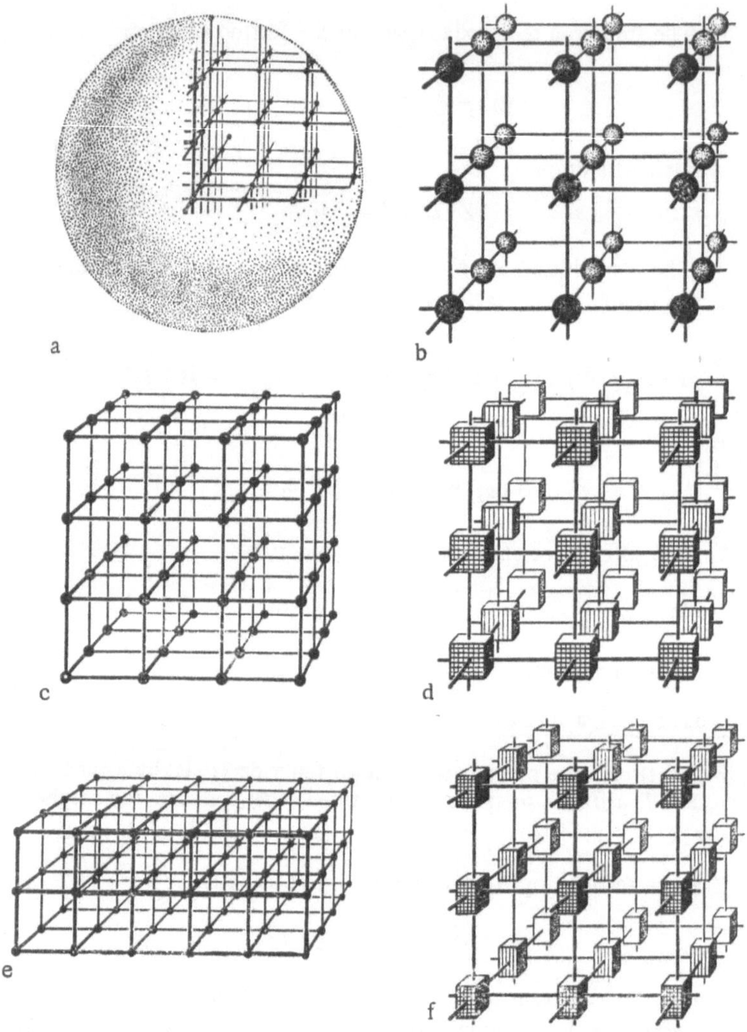

Fig. 5. The simplest direct and reciprocal lattices; a) is a cubic crystal in the form of a sphere of radius R; b) is a cubic reciprocal lattice composed of spherical nodes of radius 1/R; c) is a cubic crystal in the form of a cube of edge a; d) is a reciprocal lattice composed of small cubes with edges 1/a; e) is a crystal in the form of a right-angled parallelepiped with edges a, b, and c; f) shows the nodes of the reciprocal lattice in the shape of parallelepipeds with edges 1/a, 1/b, 1/c; g) is a two-dimen-

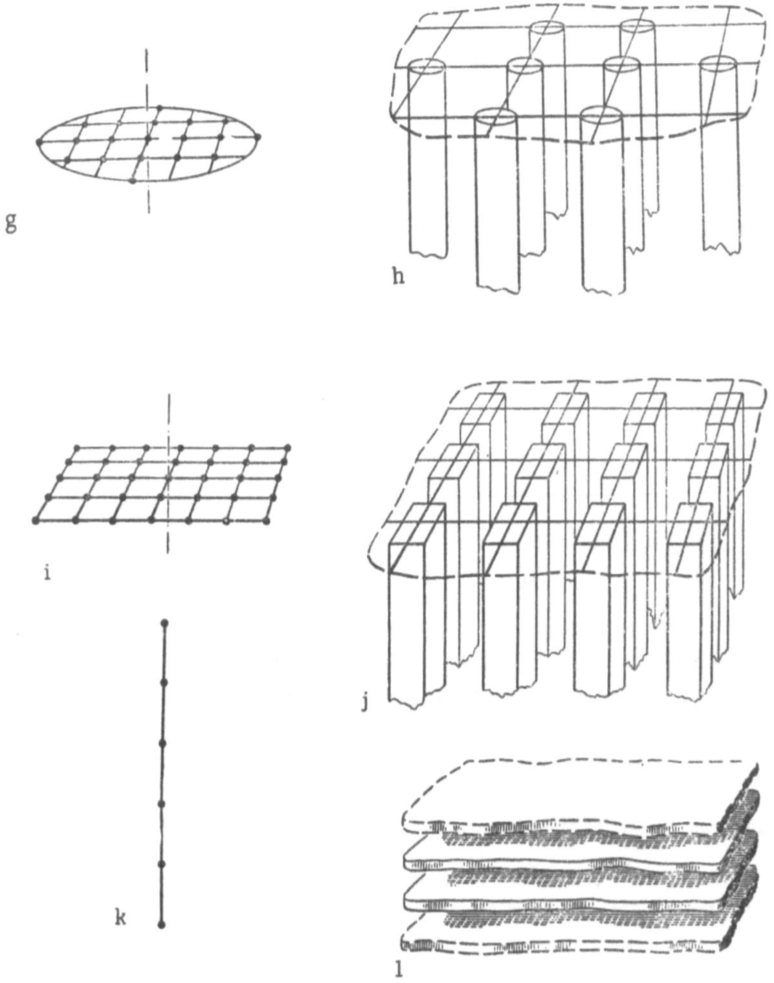

sional lattice in the form of a circle of radius R; h) is its reciprocal lattice made up of rods of radius 1/R; i) is a two-dimensional lattice in the form of a rectangle; j) is its reciprocal lattice. The rods have a cross section with dimensions 1/a, 1/b; k) is a one-dimensional lattice of length L; 1) is its reciprocal lattice consisting of planes 1/L in thickness.

whence

$$\mathfrak{F}=\frac{Z}{V}\sum s(\mathbf{H}-\mathbf{H}_{hkl}). \qquad (67)$$

Since $s(\mathbf{H}-\mathbf{H}_{hkl})$ cannot extend to a neighboring node, then

$$\mathfrak{F}^2=\frac{Z^2}{V^2}\sum |s(\mathbf{H}-\mathbf{H}_{hkl})|^2. \qquad (68)$$

It is not hard to see that the image of a finite lattice in reciprocal space is a reciprocal lattice, the nodes of which, in contrast to the case of an infinite lattice, are no longer points.

Let, for example, the finite lattice under study occupy the volume of a sphere of radius R. Then, as has been shown in Section 7, the function s(H) also has spherical symmetry

$$s(H)=\frac{4}{3}\pi R^3\Phi(2\pi HR)$$

and drops practically to zero when $H \approx 1/2R$. This means that a node of the reciprocal lattice is a sphere with the radius indicated.

The radius of this sphere is considerably smaller than the period of the reciprocal lattice, since the ratio of the volume of a cell of reciprocal space to the volume of a node in reciprocal space is equal to the ratio of the volume of the region of physical space occupied by the lattice to the volume of a unit cell of physical space. Thus, the radius of a diffuse node of the reciprocal lattice is $\sqrt[3]{N}$ times smaller than a half period of the reciprocal lattice, if N is the number of unit cells of the finite lattice.

In an overwhelming number of cases the dimensions of a node are so insignificant that a study of the intensity distribution within a node becomes impossible. In this case it has been established by experiment that the value of the integrated (over the node) intensity is

$$I_{hkl}=\frac{Z^2}{V^2}\int s^2(\mathbf{H})\,d\tau.$$

The value of this integral is independent of the shape of the object. On the basis of (I, 15), which is true for any pair of Fourier transforms, we obtain

$$\int s^2(\mathbf{H})\,d\tau=\int \varphi^2(\mathbf{R})\,dv.$$

But the last integral, according to the definition of the function $\varphi(\mathbf{R})$, is equal to the volume occupied by the lattice NV. Thus, for a finite point lattice the following expression is true:

$$I_{hkl_{\text{meas}}}=N\frac{Z^2}{V}, \tag{69}$$

i.e., the product of the square of the electron density by the volume occupied by the lattice.

The shape of the node is determined by that of the object. All the nodes of the reciprocal lattice (including the zero one) have the same shape and, in addition, the same shape that would be given by the same object with a constant (and not periodically distributed) density.

The formula just obtained (68) is true not only for a three-dimensional lattice, but for any lattice. Examples of finite lattices and their images are given in Fig. 5.

9. The Ideal Finite Crystalline Lattice[7]

In x-ray crystal analysis, it is not the real structure of the crystal that is determined, as is shown below, but the ideal crystal lattice which approximates most accurately the object under study. Therefore, the case analyzed in this paragraph is basic for the theme of the present book.

We shall understand by an ideal crystal lattice a finite region (cut out from space) of the ideally periodic lattice, the distribution of matter in the unit cell of which is given by some function $\rho_0(\mathbf{r})$ positive at all $\mathbf{r}$, equal to zero in the whole of space, except in the volume of the cell, and otherwise arbitrary.

The electron density of a finite ideal crystalline lattice is

$$\rho(\mathbf{R})=\varphi(\mathbf{R})\sum_{L}\rho_0(\mathbf{R}-\mathbf{r}_L). \tag{70}$$

We find the scattering amplitude of an object as the Fourier transform of $\rho(\mathbf{R})$

$$\mathfrak{F}=\Phi(\rho),$$

[7]The ideas for deriving the formulas of this and the preceding paragraphs by means of the theory of Fourier integrals are due to Ewald [10], and have been developed by Hoseman.

and the intensity as the Fourier transform of the self-convolution, i.e.,

$$\mathfrak{F}^2 = \Phi\left[S(\rho^i, \rho)\right].$$

To carry out these computations one should note that the electron density of a finite ideal crystal can be represented as the convolution of the product of the functions $\frac{1}{Z}\varphi(\mathbf{r})\,\mathfrak{G}(\mathbf{r})$ with the function $\rho_0(\mathbf{r})$.

In fact,

$$\rho(\mathbf{r}) = \frac{1}{V}\int \varphi(\mathbf{r}') \sum \delta(\mathbf{r}' - \mathbf{r}_L)\rho_0(\mathbf{r} - \mathbf{r}')\,dv',$$

which is the same as (70). Thus

$$\rho = S\left(\frac{\varphi\mathfrak{G}}{Z},\ \rho_0\right). \tag{71}$$

Then the scattering amplitude is equal to

$$\mathfrak{F} = \Phi\left[S\left(\frac{\varphi\mathfrak{G}}{Z},\ \rho_0\right)\right] = \Phi\left(\frac{\varphi\mathfrak{G}}{Z}\right)\Phi(\rho_0).$$

The Fourier transform of $\rho_0(\mathbf{r})$ is nothing else but the scattering amplitude of the unit cell of the crystal. We shall call this quantity the structure amplitude of the cell:

$$F(\mathbf{H}) = \int \rho_0(\mathbf{r})\,e^{i2\pi\mathbf{H}\mathbf{r}}\,dv. \tag{72}$$

The Fourier transform of $\varphi\,\mathfrak{G}$ is found in the preceding paragraph; it is equal to

$$\frac{Z}{V}\sum s(\mathbf{H} - \mathbf{H}_{hkl}).$$

Thus, the scattering amplitude of a finite ideal crystal lattice is given by the expression

$$\mathfrak{F}(\mathbf{H}) = \frac{F(\mathbf{H})}{V}\sum s(\mathbf{H} - \mathbf{H}_{hkl}). \tag{73}$$

For the reasons mentioned when writing (68),

$$\mathfrak{F}^2(\mathbf{H}) = \frac{F^2}{V^2}\sum |s(\mathbf{H}-\mathbf{H}_{hkl})|^2. \tag{74}$$

For the integrated intensity of each reflection all the arguments of the preceding paragraph are also true. Consequently,

$$I_{hkl} = \frac{N}{V}F^2.$$

We see that experimentally the value of F^2 is obtained only at the nodes of reciprocal space.

Thus, assuming the kinematic diffraction theory to be true, and knowing the characteristic angular factor for the photographic method, we obtain experimentally the values of the integrated intensities of the spots I_{hkl}, and convert them by means of the last formula into moduli of structure amplitudes, i.e., into moduli of the scattering amplitude of the unit cell.

To pass over from $|F_{hkl}|$ to the structure of the crystal two methods exist. The first is the direct finding of the electron density ρ_0; the second is the finding of the self-convolution of the electron density $S(\rho_0, \rho_0)$ with the subsequent determination of ρ. It is convenient to compute, not these functions themselves, but their periodic extensions into infinity; in other words – the convolutions of these functions with the lattice function

$$\rho_\infty(\mathbf{r}) = S\left(\rho_0, \frac{1}{Z}\mathfrak{G}\right). \tag{75}$$

The correctness of this is evident. For example,

$$\begin{aligned} S\left(\rho_0, \frac{\mathfrak{G}}{Z}\right) &= \frac{1}{V}\int \rho_0(\mathbf{r}')\sum_L \delta[(\mathbf{r}-\mathbf{r}_L)-\mathbf{r}']\,dv' = \\ &= \sum \rho_0(\mathbf{r}-\mathbf{r}_L) = \rho_\infty(\mathbf{r}), \end{aligned} \tag{76}$$

and since $\rho_0(\mathbf{r})$ by definition differs from zero only within the limits of a cell, the sum just obtained is the periodic function $\rho_\infty(\mathbf{r})$.

Using the convolution formula, we obtain

$$\rho_\infty(\mathbf{r}) = S\left(\rho_0, \frac{\mathfrak{G}}{Z}\right) = \Phi^{-1}[F\cdot\Sigma\delta(\mathbf{H}-\mathbf{H}_{hkl})],$$

but

$$\Phi^{-1}[F\Sigma\delta(\mathbf{H}-\mathbf{H}_{hkl})]=\int F\Sigma\delta(\mathbf{H}-\mathbf{H}_{hkl})\,e^{-2\pi i\mathbf{H}\mathbf{r}}\,d\tau=$$

$$=\frac{1}{V}\sum F_{hkl}e^{-2\pi i\mathbf{H}_{hkl}\mathbf{r}}.$$

Thus, the first working formula acquires the form

$$\rho_{\infty}(\mathbf{r})=\frac{1}{V}\sum F_{hkl}e^{-2\pi i\mathbf{H}_{hkl}\mathbf{r}}. \tag{77}$$

For the Patterson[8] function we obtain analogously:

$$P_{\infty}(\mathbf{r})=\Phi^{-1}[F^{2}\delta(\mathbf{H}-\mathbf{H}_{hkl})],$$

i.e.,

$$P_{\infty}(\mathbf{r})=\frac{1}{V}\sum F_{hkl}^{2}e^{-2\pi i\mathbf{H}_{hkl}\mathbf{r}} \tag{78}$$

We shall also call (78) the F^2 series.

10. The Real Crystal

The time-average electron density of the crystal is (strictly speaking) not a periodic function. Mosaic blocks, cracks, cavities, and other structural irregularities force us to write the electron density formula of the crystal in the form

$$\rho(\mathbf{R})=\varphi(\mathbf{R})\left[\sum_{L}\rho_{0}(\mathbf{R}-\mathbf{r}_{L})+\Delta(\mathbf{R})\right], \tag{79}$$

where $\Delta(\mathbf{R})$ is some function describing the deviation of the electron density from the ideal one. Evidently $\sum_{L}\rho_{0}(\mathbf{R}-\mathbf{r}_{L})$ can be considered not only as the average over time of the electron density of the whole crystal, but even as the time average in each of the cells of the crystal, if the average over all cells for every $\mathbf{r}=\mathbf{R}-\mathbf{r}_L$ is

$$\overline{\Delta(\mathbf{R}-\mathbf{r}_{L})}=0. \tag{80}$$

[8]Formula (78) was first published by Patterson in 1935.

The task of structure analysis is primarily to find the electron density of an ideal crystal $\rho = \sum \rho_0 (\mathbf{R} - \mathbf{r}_L)$, which approximates $\bar{\rho}$ as closely as possible. However, a knowledge of $\Delta(\mathbf{R})$ is also important for a number of problems.

The scattering amplitude of a real crystal is $\mathfrak{F} = \Phi(\rho) + \Phi(\Delta)$. The intensity found experimentally is equal to

$$\mathfrak{F}^2 = \Phi\{S[(\rho + \Delta)', (\rho + \Delta)]\}, \tag{81}$$

but

$$\begin{aligned} S[(\rho + \Delta)', (\rho + \Delta)] = \int \rho(\mathbf{R}')\rho(\mathbf{R} + \mathbf{R}')\,dv' + \\ + \int \Delta(\mathbf{R}')\Delta(\mathbf{R} + \mathbf{R}')\,dv', \end{aligned} \tag{82}$$

since

$$\int \rho(\mathbf{R}')\Delta(\mathbf{R} + \mathbf{R}')\,dv' = 0.$$

Substituting (82) into (81) we see that the intensity $\mathfrak{F}^2(\mathbf{H})$ is the sum of two terms, the first of which is the scattering intensity of an ideal crystal of density ρ, i.e., (81) can be represented in the form

$$\mathfrak{F}^2 = \frac{F^2}{V^2} \sum |s(\mathbf{H} - \mathbf{H}_{hkl})|^2 + I', \tag{83}$$

where I' is the scattering due to defects, thermal waves, etc.

As we have seen in the preceding paragraph, the first term in (83) differs from zero only at the nodes of the reciprocal lattice. The second term can, generally speaking, reach considerable magnitudes anywhere in reciprocal space.

To separate out I' in (83), we always use the following, generally arbitrary, assumption: the value of I' within the node is assumed to be equal to the values of I' in the vicinity of the node. Under this condition, the first term in (83) is measured as the intensity above background at a node of the reciprocal lattice.

As must be clear from comparing this section with section 1, scattering enters into I' from thermal vibrations as well as from static lattice distortions.

These two effects can be separated in one way only: by taking measurements at different temperatures.

11. Structure Amplitudes and Products

It is possible to obtain by well known methods ([4] ch. IV) values of the structure amplitudes F for all the nodes of the reciprocal lattice from the intensities of the x-ray reflections (taking into account the reservations made in the preceding paragraph). The structure amplitude F_{hkl} or $F_{\mathbf{H}}$ (we shall often use such an abbreviated notation, meaning by **H** a triad of indices) is equal by definition to

$$F_{\mathbf{H}} = \int \rho_0(\mathbf{r})\, e^{2\pi i \mathbf{H}\mathbf{r}} dv.$$

As is evident from Section 4 (see also below, chapter VI), it is possible to substitute, with an accuracy quite sufficient for the purpose of structure analysis, $\rho_0(\mathbf{r})$ by a sum of spherically symmetrical atomic functions:

$$\rho_0(\mathbf{r}) = \sum_{j=1}^{N} \rho_j(\mathbf{r} - \mathbf{r}_j),$$

where N is the number of atoms in the unit cell.

Substituting into the preceding formula and using the definition (28) of the atomic factor, we obtain

$$F_{\mathbf{H}} = \sum_{j=1}^{N} f_j e^{2\pi i \mathbf{H}\mathbf{r}_j} \tag{84}$$

or, for a centrosymmetric crystal,

$$F_{\mathbf{H}} = 2 \sum_{j=1}^{N/2} f_j \cos 2\pi \mathbf{H}\mathbf{r}_j .$$

Further simplification is achieved by introducing a unitary atomic factor with the aid of which the structure amplitude takes the form

$$F_{\mathbf{H}} = \hat{f} \sum_{j=1}^{N} Z_j e^{2\pi i \mathbf{H}\mathbf{r}_j}. \tag{85}$$

Dividing both sides of the last equation by $Z\hat{f}$, we come to the concept of the unitary structure amplitude first introduced by Harker and Kasper [11]:

$$\hat{F}_{\mathbf{H}}=\sum_{j=1}^{N} n_j e^{2\pi i \mathbf{H}\mathbf{r}_j}, \tag{86}$$

where $n_j = \frac{Z_j}{Z}$.

The maximum value of $\hat{F}$ (when all the atoms of the cell are scattering in phase) is equal to unity—hence the nomenclature.

As will be apparent from what comes later, it is expedient for purposes of structure analysis to group unitary structure amplitudes into so-called structure products. The expression

$$\hat{F}_{\mathbf{H}}\hat{F}_{\mathbf{K}}\hat{F}_{\mathbf{L}}\ldots\hat{F}_{\mathbf{H}+\mathbf{K}+\mathbf{L}+\ldots} \tag{87}$$

is called a structure product of the _n_th order [12] (according to the number of cofactors).

A product can be degenerate, if $\mathbf{H} = \mathbf{K}$ or $\mathbf{H} = \mathbf{K} = \mathbf{L}$, and so forth. If the p's of the amplitudes are the same, then this structure product will be called p-fold degenerate. Let us give examples:

product of the third order:

undegenerate $\hat{F}_{\mathbf{H}}\hat{F}_{\mathbf{K}}\hat{F}_{\mathbf{H}+\mathbf{K}}$,

degenerate $\hat{F}^2_{\mathbf{H}}\hat{F}_{2\mathbf{H}}$,

product of the fourth order:

undegenerate $\hat{F}_{\mathbf{H}}\hat{F}_{\mathbf{K}}\hat{F}_{\mathbf{L}}\hat{F}_{\mathbf{H}+\mathbf{K}+\mathbf{L}}$,

degenerate $\hat{F}^2_{\mathbf{H}}\hat{F}_{\mathbf{K}}\hat{F}_{2\mathbf{H}+\mathbf{K}}$,

product of the fifth order:

undegenerate $\hat{F}_{\mathbf{H}}\hat{F}_{\mathbf{K}}\hat{F}_{\mathbf{L}}\hat{F}_{\mathbf{M}}\hat{F}_{\mathbf{H}+\mathbf{K}+\mathbf{L}+\mathbf{M}}$,

simply degenerate $\hat{F}^2_{\mathbf{H}}\hat{F}_{\mathbf{K}}\hat{F}_{\mathbf{L}}\hat{F}_{2\mathbf{H}+\mathbf{K}+\mathbf{L}}$,

two-fold degenerate $\hat{F}^2_{\mathbf{H}}\hat{F}^2_{\mathbf{K}}\hat{F}_{2\mathbf{H}+2\mathbf{K}}$

etc.

Structure products of the third order $\hat{F}_{\mathbf{H}}\hat{F}_{\mathbf{K}}\hat{F}_{\mathbf{H}+\mathbf{K}}$ are of basic importance in the problem that interests us. We shall call them simply structure products when this does not lead to confusion.

Written out fully, the formula of this structure product for a centrosymmetric crystal has the form

$$X_{\mathbf{HK}}=\hat{F}_{\mathbf{H}}\hat{F}_{\mathbf{K}}\hat{F}_{\mathbf{H}+\mathbf{K}}=2^3\sum_{j,m,p=1}^{N/2} n_j n_m n_p \cos\alpha_j \cos\beta_m \cos(\alpha_p+\beta_p), \tag{88}$$

where $\alpha = 2\pi\mathbf{H}\mathbf{r}$ and $\beta = 2\pi\mathbf{K}\mathbf{r}$.

Finally, let us introduce the following definitions: the expression $n_j \exp 2\pi i \mathbf{H}\mathbf{r}$ will be called a component of a (unitary) structure amplitude; the expression $n_j n_m n_p \exp 2\pi i \mathbf{H}\mathbf{r}_j \exp 2\pi i \mathbf{K}\mathbf{r}_m \cdot \exp 2\pi i (\mathbf{H} + \mathbf{K}) \mathbf{r}_p$ will be called a component of a structure product (in this case of the third order).

12. A Comparison of X-ray, Electron, and Neutron Structure Analyses

The experimental procedure in these three diffraction methods is substantially different. However, the methods of treating the results of observation are almost identical. This is due to the complete analogy between the formulas which relate the distribution of the scattering intensity to the structure.

It is possible to compute by analogous methods the electron density of a substance from the x-ray diffraction intensity data, the electrostatic potential from the electron diffraction data, and, finally, the distribution of atomic nuclei from the neutron diffraction data.

It is clear that all diffraction methods complement one another, and their joint application to the solution of the structure of one and the same object promises to be very interesting.

Now we shall analyze the specific properties of neutron and electron diffraction, and we shall show that the intensities of the diffraction spots on electron and neutron patterns can be used to compute functions which give the "structure" of the object.

If the amplitudes of the x-rays, electrons, and neutrons scattered by some object are $\mathfrak{F}_x$, $\mathfrak{F}_e$, and $\mathfrak{F}_n$, and the electron density, electrostatic potential, and nuclear "density" are $\rho(\mathbf{r})$, $\varphi(\mathbf{r})$, and $n(\mathbf{r})$, then the connection between the experimental data and the structure is expressed by formulas similar to Fourier transforms:

$$\rho = \Phi(\mathfrak{F}_x); \quad \varphi = \Phi(\mathfrak{F}_e); \quad n = \Phi(\mathfrak{F}_n). \tag{89}$$

If the phases of the amplitudes are unknown, then the self-convolutions of these structural functions may be constructed directly from experiment.

$$\Phi^{-1}(\mathfrak{F}_x^2) = S(\rho', \rho); \quad \Phi(\mathfrak{F}_e^2) = S(\varphi', \varphi); \quad \Phi(\mathfrak{F}_n^2) = S(n', n).$$

Let us examine some details.

Diffraction of Electrons.[9] The motion of electrons is described by Schrödinger's wave equation,

$$\nabla^2\Psi + \frac{8\pi^2 m}{h^2}(E - V)\psi = 0, \tag{90}$$

where ψ (xyz) is the wave function, the squared modulus of which gives the probability of finding an electron at a given point. The full energy E of the electron beam is set by the accelerating potential p: E = ep (e is the electronic charge) and determines the wavelength of the monochromatic wave incident on the object:

$$\psi_0 = Ae^{ik_0 x}, \tag{91}$$

so that $\lambda^{-1} = k_0/2\pi = \sqrt{2mE}/h$; ψ_0 is a solution of (90) in the absence of the term V (potential energy), which becomes important only when the wave impinges on some object or other – an atom, a molecule, a crystal, in the potential φ (**r**) of which the electron acquires the potential energy $V(\mathbf{r}) = \varphi(\mathbf{r})e$. This is what causes scattering. Thus the "scattering material" in electron diffraction is the electrostatic potential φ (**r**), which plays the same part here as the electron density ρ (**r**) in the diffraction of x-rays. This analogy, as we shall see later, is complete and extends to the basic formulas of the theory of scattering.

The solution of Schrödinger's equation in the kinematic approximation i.e., assuming the weakness of the secondary beams, is carried out by representing the sought function ψ as the sum of the initial wave ψ_0 and of the small scattered wave ψ'

$$\psi = \psi_0 + \psi'.$$

Substituting this expression in (90), and keeping in mind that ψ_0 is the solution of (90) without the term V, we obtain for ψ' the equation

$$\nabla^2\psi' + k_0^2\psi' = U(\mathbf{r})\,\psi(\mathbf{r});\quad U(\mathbf{r}) = \frac{8\pi^2 me}{h^2}\,\varphi(\mathbf{r}). \tag{92}$$

In mathematical form it is a so-called generalized Poisson equation, which describes, in particular, the propagation of an electromagnetic disturbance in an environment containing charges – which confirms anew the above mentioned analogy. Its solution

$$\psi'(\mathbf{r}) = \frac{1}{4\pi}\int U(\mathbf{r}_1)\,\psi(\mathbf{r}_1)\frac{e^{ikR}}{R}\,dv_1, \tag{93}$$

[9]This and the following paragraphs are written by B. K. Vainshtein.

where $R = |\mathbf{r} - \mathbf{r}_1|$ ($\mathbf{r}_1$ is a vector inside the scattering volume).

Here $\psi = \psi_0 + \psi'$ is under the integral sign. Using the basic condition of kinematic scattering – the weakness of the secondary waves – $\psi' \ll \psi_0$, i.e., substituting ψ by ψ_0, which means neglecting the further scattering of the secondary waves ψ' which have arisen, we obtain

$$\varphi' = \frac{1}{4\pi}\int U(\mathbf{r})\, Ae^{i\mathbf{k}_0\mathbf{r}}\,\frac{e^{ikR}}{R}\,dv. \tag{94}$$

The same kind of term is neglected in the kinematic theory of x-ray scattering. In this case, the assumption is even more justified, as the absolute magnitude of the amplitudes of the secondary x-ray waves is much smaller than that of electrons, which react with the object much more strongly. Refusing to neglect this kind of term is the basic feature of the dynamic theory of x-ray and electron scattering. Substituting in (94)[10] R for $r - (\mathbf{n}\mathbf{r}_1)$ ($\mathbf{n}$ is the unit vector of the direction $\mathbf{k}$) so that $e^{ikR} = e^{ikr}e^{i\mathbf{k}\mathbf{r}_1}$, and keeping in mind that $\mathbf{k}_0 - \mathbf{k} = 2\pi\mathbf{H}$, i.e., introducing a vector in reciprocal space, we find

$$\psi' = \frac{2\pi me}{h^2}\,\frac{e^{ikr}}{r}\,A\mathfrak{F}_{el}(\mathbf{H}). \tag{95}$$

This means that at a great distance from the object, ψ' is a spherical wave with an amplitude proportional to the original amplitude A and to the Fourier integral $\mathfrak{F}(\mathbf{H})$ (calculated from the potential of the object):

$$\mathfrak{F}(\mathbf{H}) = \int \varphi(\mathbf{r})\, e^{2\pi i\mathbf{H}\mathbf{r}}\,dv. \tag{96}$$

The factor $\frac{2\pi me}{h^2}$ is analogous to the factor $\frac{e}{mc^2}$ in x-ray diffraction, and can be considered as the scattering by some unit of potential similar to that in x-ray diffraction where scattering is expressed in electron units.

If an isolated atom is considered to be the scattering object, then, converting (96) to spherical coordinates, we obtain an expression for the atomic scattering amplitude:

$$f_{el}(H) = \int \varphi(r)\, r^2\,\frac{\sin 2\pi Hr}{2\pi Hr}\,dr, \tag{97}$$

[10] This equation is correct for sufficiently large distances from the scattering volume under the condition that $\mathbf{k} \parallel \mathbf{r}$.

which is quite analogous to the expression for the x-ray atomic factor. Just as in Section 10, where analyzing the scattering from a crystal $\mathfrak{F}_{el}$ may be expressed by (73) by introducing the structure amplitude of the cell $F_{el}(\mathbf{H})$.

Representing the potential of the cell as the sum of the potentials of its component atoms

$$\varphi(\mathbf{r}) = \sum_i \varphi(\mathbf{r} - \mathbf{r}_i),$$

the structure amplitude may be represented in the usual form as a sum of atomic amplitudes with phase factors

$$F_{el}(\mathbf{H}) = \sum f_{el}\, e^{2\pi i(\mathbf{r}_i\mathbf{H})}. \tag{98}$$

Analogously to (77) we obtain a representation of the potential $\varphi(xyz)$ in the crystal by the Fourier series

$$\varphi(xyz) = \frac{1}{V} \sum \sum \sum F_{el}\, e^{-2\pi i(\mathbf{r}\mathbf{H})}.$$

The moduli $|F_{el}|$, as in x-ray diffraction, are found from the intensities of the reflections on electron diffraction patterns. Thus the refinement of the experimental electron diffraction data by the Fourier method makes it possible to obtain a picture of the potential [13, 14] φ (xyz) of the crystalline lattice.

The complete analogy of the mathematical apparatus makes the theory of determining crystal structures and the practical methods of interpretation (for example, F^2_{el} series, etc.) general for x-rays, electrons, and neutrons. Nevertheless, differences in the physical nature and the character of the functions which determine the scattering lead to some physical differences between the methods.

Let us point out the basic peculiarities of the potential function $\varphi(xyz)$.

The potential of a crystal is composed of the positive potential of the nuclei $\varphi_+ = \dfrac{Ze}{r}$ and the negative potential of the electron shells φ_-:

$$\varphi = \varphi_+ - \varphi_-.$$

As a consequence of the concentrated character of the nuclear charges and the diffuseness of the electron shells $\varphi_+ > \varphi_-$, i.e., $\varphi > 0$. The importance of the electron shells in screening the nuclear potential, and the shape of the atomic potential is affected not only by the magnitude, but also by the character of the electron distribution in the shells. The potential of the crystal, made up of a superposition of the atomic potentials, is a positive, continuous periodic function, the maxima of which correspond to the atoms.

The potential φ (xyz) is a more diffuse function than the electron density ρ (xyz), as follows from a well known consequence of the Thomas-Fermi statistical theory of the atom:

$$\rho \sim \varphi^{3/2}; \quad \varphi \sim \rho^{2/3}. \tag{99}$$

This is shown schematically in Fig. 6. From this it follows that there is a more rapid diminution of the curves of atomic scattering for electrons f_{el} as compared with those for x-rays f_x. At the same time it is apparent from (99) and from Fig. 6 that the ratios of the heights of the peaks of light and heavy atoms will differ in x-ray and in electron diffraction: in electron diffraction this ratio is more advantageous for the light atoms.

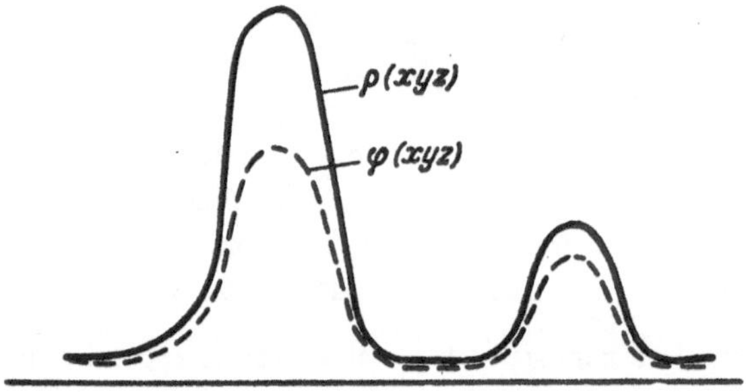

Fig. 6. A comparison of the shape of the electron density ρ (xyz) and the potential φ (xyz). The peaks of electron density are sharper, but the ratio of the heights of the peaks in φ is smaller than that of the peaks in ρ.

The distinguishing property of F_{el}- and F^2_{el}- series in electron diffraction is that they have about two to four times fewer terms than in the case of x-ray diffraction, which is a consequence of the more rapid decline of the atomic scattering curve. For the same reason the electron diffraction series converge more rapidly than the x-ray diffraction ones, the termination error is smaller in them, and the function obtained is less sharp. In conclusion, let us point out that inasmuch as the distribution of the potential and charges in a crystal are connected by Poisson equation

$$\nabla^2\varphi = -4\pi(\rho_+ - \rho_-), \tag{100}$$

the structure amplitudes for electrons F_{el} and for x-rays F_X turn out also to be connected by the relation

$$F_{el} = \frac{F' - F_X}{\pi H^2}, \tag{101}$$

where $F' = \sum_j Z_j e^{2\pi i(\mathbf{r}_j\mathbf{H})}$ (Z is the nuclear charge).

Neutron Diffraction. Schrödinger's equation (90) holds not only for electrons but also for microparticles in general; in particular, it makes possible the analysis of neutron scattering. Let us turn first to the scattering of a monochromatic neutron beam by an isolated nucleus. The initial wave ψ_0 has here the same form as (91). The wavelength of the slow (thermal) neutrons used in structural neutron diffraction is about equal to 1 A. On striking a nucleus a neutron undergoes scattering which is determined in form and character from (90) by the potential function V(r) of the nucleus, by the initial energy E of the neutron, and by the mass m of the system. Analogously to the scattering of electrons, which is dependent on the electrostatic potential of the object, the scattering of neutrons is determined by the potential of the nuclear forces. The characteristic of these forces is their extremely rapid diminution with distance. The nuclear field is concentrated in the immediate neighborhood of the nucleus and has a radius of $r_0 \sim 10^{-13}$ cm. The solution again has the form of a sum of the initial and scattered waves, which, quite analogously to the case of electron diffraction, is spherically symmetrical:

$$\psi = \psi_0 + f_n(H)\frac{e^{ikr}}{r} \tag{102}$$

[see (91) and (95)]. The exact procedure for solving (90) for neutrons by substituting (102) into it leads, as is known, to a very important and very simple result, which we shall now obtain qualitatively from general considerations. As in the theory of x-ray scattering, and also in that of electron scattering, the dependence of the atomic scattering amplitude on the angle ϑ (or on H) arises as a consequence of taking into account the phase correlations between the secondary waves in the Fourier integral (5) or (96). The magnitude of the scattering volume is of the same order as the length of the impinging wave (Fig. 7a), and the secondary waves emitted from its various points have different phases. However, in the case of the scattering of slow neutrons $\lambda \gg r_0$ (Fig. 7b) and the phase difference is very small, i.e., all the secondary waves originating from a nucleus have the same phase for scattering in any direction, and therefore the scattering of neutrons by a nucleus is spherically symmetrical, and the amplitude $f_n(\vartheta)$ is not affected by the scattering angle ϑ – it is a constant magnitude. Consequently, the f_n- "curve" does not fall. Thus, for the scattering of neutrons, the ideal case of scattering from "point" atoms is realized. The structure amplitude of the coherent scattering from a crystal will have here a form analogous to (84) and (98):

$$F_{\mathbf{H}} = \sum_i f_{n_i} e^{2\pi i (\mathbf{r}_j \mathbf{H})}, \tag{103}$$

in addition the nuclear amplitude f_n here, in contrast with f_x and f_{el}, is, as we have just clarified, a constant quantity. Formally, in this case too, it would be possible to introduce the Fourier integral of the nuclear scattering power of the entire cell, representing this quantity as a sum of δ functions with weights f_n, each of which describes a point scattering center.

$$n(\mathbf{r}) = \sum f_{n_j} \delta_j (\mathbf{r} - \mathbf{r}_j),$$
$$F_{\mathbf{H}} = \int n(\mathbf{r})\, e^{2\pi i (\mathbf{r}\mathbf{H})} dv. \tag{104}$$

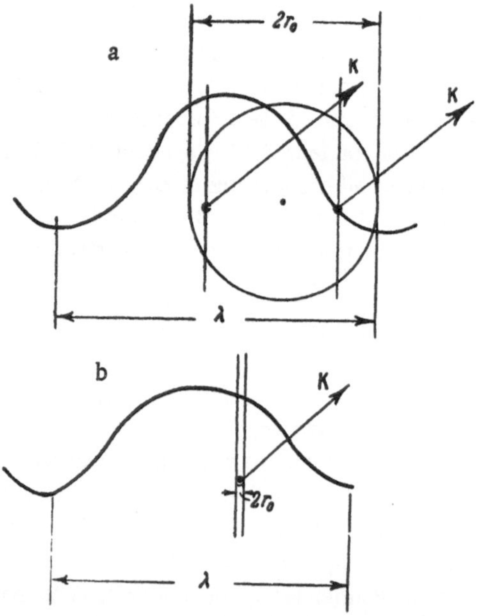

Fig. 7. Illustration of the characteristic property of the atomic scattering factor of neutrons.

However, this representation acquires a real physical meaning only when the thermal motion of the nuclei of the atoms in the crystal is taken into account. Then a point nucleus is spread over some region κ (**r**), in agreement with which the constant f_n must be multiplied in the usual manner by a temperature factor, and so diminishes with increasing sin ϑ/λ according to the law of this temperature factor. For example, for a spherically symmetrical vibration f_n = const $e^{-B\left(\frac{\sin\vartheta}{\lambda}\right)^2}$.

The superposition of diffuse nuclei with thermal motion will appear as a continuous function

$$n(\mathbf{r}) = \sum_j f_n \varkappa_i(\mathbf{r} - \mathbf{r}_j)$$

a function of the nuclear scattering power, or in short, but not quite correctly, "of the nuclear density" (not correctly, because this function has no relation whatsoever to a real density). This function is also periodic, as are also ρ (**r**) and φ (**r**). Inverting the Fourier integral (104), we obtain a representation of n (**r**) as a Fourier series:

$$n(\mathbf{r}) = \frac{1}{V} \sum \sum \sum F_{\mathbf{H}} e^{-2\pi i(\mathbf{rH})}, \tag{105}$$

where the $|F_{\mathbf{H}}|$ are found from the neutron-diffraction experiment. Thus, also, in analyzing crystals by the neutron-diffraction method, the experimental data that are obtained can be refined by the Fourier method [15]. The mathematical apparatus of all three diffraction methods is identical.

However, as in the case of the diffraction of electrons, the difference in the physical nature of the scattering carries with it a number of specific properties of the Fourier series in neutron diffraction. The most important of these is that some nuclei have positive, and some negative scattering amplitudes f_n (the majority of the f_n are positive). For example, for potassium $f_{nK} = 0.35 \cdot 10^{-12}$ cm, for carbon $f_{nC} = 0.64 \cdot 10^{-12}$ cm, and for hydrogen $f_{nH} = -0.38 \cdot 10^{-12}$ cm. If there are nuclei with negative f_n in the crystal, then they will be represented on the pattern of "nuclear density" not as peaks but as hollows. In these cases the general condition of the nonnegativity of scattering power, correct for x-rays and electrons, is violated. The values of f_n are not affected systematically by the atomic number. This allows one, for example, to determine easily a hydrogen atom in the presence of the heaviest atoms (in the usual sense) by neutron diffraction, to analyze structures which consist of atoms indistinguishable by means of x-rays and electrons because of the proximity of their atomic numbers (for example, alloys of the Fe-Co type and others), and to study phenomena connected with the differentiation of isotopic composition. The presence of a magnetic moment in the neutron makes it possible to study the magnetic structures of crystals (for example, ferromagnetics and antiferromagnetics). The basic characteristic of the Fourier series in neutron diffraction is the sharpness of the peaks – the function of "nuclear density" $n(\mathbf{r})$ is sharper than either $\rho(\mathbf{r})$ or $\varphi(\mathbf{r})$. However, because of the slow diminution of amplitudes, which are affected only by the temperature factor, termination effects are very noticeable in neutron-diffraction series.

CHAPTER III

STRUCTURE AMPLITUDES AND PRODUCTS AS RANDOM QUANTITIES

1. Statement of the Problem

Wilson [16, 17] was the first to consider the expediency of considering a structure amplitude as a random quantity, and the possibility of determining a priori its distribution functions. This assertion is true for other structure functions as well, for example, for the structure products. The meaning of such assertions is brought out in the following.

Depending on the space group of the crystal, the structure amplitudes, products, or other analogous functions may always be divided into a limited number of classes, such that within a class their formulas are identical.

We say that for the group $P2_1/a$ for all $\hat{F}_{hkl}$ for which h + k = 2n, the following equation is true:

$$\hat{F}_{hkl} = 4\sum_{j=1}^{N/4} n_j \cos 2\pi (hx_j + lz_j) \cos 2\pi k y_j; \qquad (1)$$

while, if h + k = 2n + 1

$$\hat{F}_{hkl} = -4\sum_{j=1}^{N/4} n_j \sin 2\pi (hx_j + lz_j) \sin 2\pi k y_j; \qquad (2)$$

and for the zonal reflections $h0l$

$$\hat{F}_{h0l} = 4\sum_{j=1}^{N/4} n_j \cos 2\pi (hx_j + lz_j). \qquad (3)$$

The structure products for the last group can also be divided into classes. For the product of the third order $\hat{F}_{\mathbf{H}}\hat{F}_{\mathbf{K}}\hat{F}_{\mathbf{H+K}}$ we obtain in the case when $\hat{F}_{\mathbf{H}}$ and $\hat{F}_{\mathbf{K}}$ belong to (1):

$$\hat{F}_{\mathbf{H}}\hat{F}_{\mathbf{K}}\hat{F}_{\mathbf{H+K}} = 4^3 \sum_{j,r,p}^{N/4} n_j n_r n_p \cos 2\pi (h_1 x_j + l_1 z_j) \cos 2\pi k_1 y_j \times$$
$$\times \cos 2\pi (h_2 x_r + l_2 z_r) \cos 2\pi k_2 y_r \cos 2\pi [(h_1 + h_2) x_p + (l_1 + l_2) z_p] \times$$
$$\times \cos 2\pi (k_1 + k_2) y_p. \qquad (4)$$

But if $\hat{F}_{\mathbf{H}}$ and $\hat{F}_{\mathbf{K}}$ belong to (2), then

$$\hat{F}_{\mathbf{H}}\hat{F}_{\mathbf{K}}\hat{F}_{\mathbf{H+K}} = 4^3 \sum_{j,r,p}^{N/4} n_j n_r n_p \sin 2\pi (h_1 x_j + l_1 z_j) \sin 2\pi k_1 y_j \times$$
$$\times \sin 2\pi (h_2 x_r + l_2 z_r) \sin 2\pi k_2 y_r \cos 2\pi [(h_1 + h_2) x_p + (l_1 + l_2) z_p] \times$$
$$\times \cos 2\pi (k_1 + k_2) y_p \qquad (5)$$

and so forth.

In this way, each class of structure amplitudes can be represented in the general form

$$\hat{F}_{\mathbf{H}} = n \sum_{j=1}^{N/n} n_j \psi (\mathbf{H}, \mathbf{r}_j), \qquad (6)$$

where ψ $(\mathbf{H}, \mathbf{r})$ is the trigonometric part of the structure amplitude, i.e., a function of the vector $\mathbf{H}$ (the node of the reciprocal lattice), and of the vectors $\mathbf{r}_j$ – the atomic centers in the cell.

The structure product of the $\underline{m}$th order can also be given for each class by the general formula

$$X_{\mathbf{H},\,\mathbf{K},\,\mathbf{L},\ldots} = n^m \sum_{j,r,p,\ldots}^{N/n} n_j n_r n_p \ldots n_m \psi (\mathbf{H},\ \mathbf{K},\ \mathbf{L}, \ldots,\ \mathbf{r}_j). \qquad (7)$$

Experiment gives us hundreds and thousands of structure amplitudes; millions of structure products may be formed. We are justified in posing the following question in relation to these and analogous magnitudes: How will the distribution function of the structure amplitude, of the product, etc., appear; in other words, we may pose the question as to the a priori computation of the fraction of the F-magnitudes (X-magnitudes), larger (or smaller) than some F_0 (some X_0). Taking into account the considerable volume of the experiment, we may expect good agreement of the computed probabilities with the observed frequencies, if the computation has been correctly carried out.

Analyzing (6) and (7), we see, generally speaking, that two distributions exist for these magnitudes: first, the computation may be carried out for fixed nodes of the reciprocal lattice, while the vector $\mathbf{r}_j$ runs through all the points of the unit cell; second, it is possible to calculate the distribution function of F or X, considering the vectors $\mathbf{r}_j$ invariable and forcing the vector **H** (or vectors **H**, **K**, **L**) to run through all the nodes (or all the pairs, all the triplets of nodes) of the reciprocal lattice. It stands to reason that only the second way is accessible for experimental checking.

Although h, k, l and x, y, z enter symmetrically into the argument of the trigonometric functions, only the first of these are whole numbers. If x, y, z run through all the values within the unit cell, then the magnitudes $hx + ky + lz$ will be uniformly distributed over the trigonometric circle. However, Veilem has proved a less obvious statement, namely, that the same uniform distribution over the trigonometric circle arises also in the case when h, k, l take all the whole-number values, with x, y, z fixed.[1]

This situation, very important for the theory of structure analysis, is not difficult to check by computation from examples. In Table 2 it is shown that for the three values of $\mathbf{r}_j$ sampled from anthracene, even for a limited number of nodes, a uniform distribution of $hx + ky + lz$ is already realized with good accuracy. It is evident that the more complex the structure, the more "disordered" it is, and the better the basic assumption of the statistical theory is fulfilled. On the contrary, for crystals with a small number of atoms in the cell, with heavy atoms in special positions, the basic assumption of the statistical theory may be contradicted; that is, the distribution of $hx + ky + lz$ will become nonuniform, and the distributions over the coordinates and the indices will be different.

What has been said enables us to consider structure amplitudes, products, and other analogous structure functions as sums of random magnitudes.

Each addend—component of the amplitude or the product—is a random magnitude which possesses a definite distribution law. This law can be found by using, for example, the following assumption: The argument of the trigonometric function is a random magnitude, uniformly distributed over the trigonometric circle.

In structural amplitudes all the components are identical and have one and the same distribution function. In structure products the components are divided into groups with different distribution functions (components of the general type — indices j,r,p different, components with equal indices j,r,p, etc., for details see page 119).

[1] Translator's note: x,y,z, must not be rationally related for the statement to be true.

Each component is a function of one [formula (3)], of two [formulas (1) and (2)], or of several trigonometric arguments α, β. The components of the structure product (4) or (5) are functions of four or six arguments α,β, considered as uniformly distributed random quantities.

The general statement of the problem for amplitudes and products is this: It is necessary to find the distribution function of the structure amplitude or product or of some other function of the structure W, which is the sum of a large number of magnitudes:

$$W = \sum_{k=1}^{M} w_k. \tag{8}$$

Furthermore

$$w_k = \psi_k(\alpha, \beta \ldots), \tag{9}$$

where ψ_k are the given trigonometric functions of one or several arguments uniformly distributed over the trigonometric circle. In our cases ψ_k will be either identical (the case of the structure amplitude) or will differ in degree of degeneracy (the case of the structure product).

It is always convenient to use a random quantity whose center of distribution is at zero. Let $\overline{W}$ be the average value of the structural function. Since the average of the sum is equal to the sum of the averages, then

$$U = \sum_{k=1}^{M} u_k, \tag{10}$$

where

$$U = W - \overline{W}, \quad u_k = w_k - \overline{w}_k. \tag{11}$$

Let us formulate the problem in more detail. The formula of the structure product or amplitude is under consideration. Its value is determined by the arguments of the trigonometric functions. Let us denote the arguments by α, β, γ, which differ from one another in the value of the index or of the coordinate, or both. Now we consider the magnitudes α, β, γ as independently random (this, strictly speaking, is incorrect, but we disregard the correlation between isolated arguments and consider the consequence of this premise in Chapter IV). We shall give to the magnitudes α, β, γ .... all kinds of values within the limits from 0 to 2π, filling this interval uniformly. For each test we shall compute the value of the structural function. After obtaining

TABLE 2. The Example of Applying the Statistical Theory to Anthracene

Interval of the argument in fractions of 2π	The number of nodes giving the stated value of the argument. The nodes are taken from a spherical shell (between Bragg angles $\theta = 36°$ and $\theta = 55°$)		
	r_1	r_2	r_3
0 —0.05 0.95—1	49	51	53
0.05—0.10 0.90—0.95	36	68	46
0.10—0.15 0.85—0.90	54	38	39
0.15—0.20 0.80—0.85	38	60	39
0.20—0.25 0.75—0.80	51	39	50
0.25—0.30 0.70—0.75	47	58	50
0.30—0.35 0.65—0.70	39	36	46
0.35—0.40 0.60—0.65	53	46	52
0.40—0.45 0.55—0.60	49	25	46
0.45—0.50 0.50—0.55	47	42	42

millions of figures, we shall compute the number of those that lie between 0 and 0.1, between 0.1 and 0.2, etc. The distribution function sought is a diagram in which the values of the structural function are laid off along the axis of abscissas, while the frequency with which these values occur is laid off along that of ordinates.

2. The Distribution Function of a Sum of Independent Quantities

Our task consists in finding the distribution function of the random quantity U (10). Let us recall certain definitions accepted in the theory of probability.

The integrated distribution function F (U) for a given $U = U_0$ gives the probability that $U < U_0$. In practice this means the following: Suppose that a large number p have been obtained of the random magnitude U. Then $F(U_0)$ is approximately equal to $\frac{p_0}{p}$, where p_0 is the number of values for which $U < U_0$. Evidently F(U) is a normalized function: $F(-\infty) = 0$ and $F(+\infty) = 1$.

The differential distribution function f(U) or, simply, the distribution function is the derivative of $f(U) = \frac{dF}{dU}$.

It is evident that f(U)dU gives the fraction of the values of the random magnitude U contained in the interval from U to U + dU.

To compute the distribution function of a sum of random magnitudes Liapunov introduced the characteristic distribution function Θ (y), which is determined in the following manner:

$$\Theta(y) = P(y) + iQ(y) = \int_{-\infty}^{+\infty} f(U) e^{iyU} dU. \tag{12}$$

This function possesses an important property for our problem, quite obvious from (12): if $U = \sum_{k=1}^{M} u_k$, then the product of the characteristic functions of the addends is equal to the characteristic function of the sum. In fact, if the u_k are independent, then, according to the theorem of multiplying probabilities

$$f(U)\,dU = f_1(u_1)\,f_2(u_2)\ldots\,du_1 du_2\ldots$$

and consequently

$$\Theta(y) = \int_{-\infty}^{\infty} e^{iyu_1} f_1(u_1)\,du_1 \int_{-\infty}^{+\infty} e^{iyu_2} f_2(u_2)\,du_2\ldots$$

From this follows

$$\Theta(y) = \Theta_1(y)\,\Theta_2(y)\ldots \tag{13}$$

or

$$\ln \Theta(y) = \sum_{k=1}^{M} \ln \Theta_k(y). \tag{14}$$

From (12) we see that the characteristic and distribution functions are a pair of Fourier transforms. Therefore the distribution function of the sum may be computed from the formula

$$f(U) = \frac{1}{2\pi} \int_{-\infty}^{+\infty} e^{-iyU} \Theta(y)\,dy, \tag{15}$$

or, since f(U) is a real function,

$$f(U) = \frac{1}{\pi} \int_{0}^{+\infty} P(y) \cos Uy dy + \frac{1}{\pi} \int_{0}^{\infty} Q(y) \sin Uy\,dy. \tag{16}$$

Thus, in principle, the problem[1] is solved, since the distribution of the addends, and consequently, of the characteristic functions $\Theta_k(y)$ also are assumed to have been derived.

Knowing $\psi_k(\alpha, \beta\ldots)$ in (9) we can compute the distribution function of each kth addend. However, there is no necessity to do this. To compute the characteristic function $\Theta_k(y)$ necessary for the computation of (14) and then (15), we expand it into a Maclaurin series:

[1]We gave a formal proof of (16). The reader will find the strict mathematical basis in specialized writings on the theory of probabilities (see, e.g., [18] p. 400 ff.).

$$\Theta(y) = 1 + \sum_{n=1}^{\infty} \frac{1}{n!} [\Theta^{(n)}(y)]_{y=0}\, y^n. \tag{17}$$

Let us analyze now what the nth derivative of the characteristic function represents at the zero point. From the definition (12), differentiating under the integral sign, we obtain

$$\Theta'(y) = i \int_{-\infty}^{+\infty} U e^{iUy} f(U)\, dU;$$

$$\Theta''(y) = - \int_{-\infty}^{+\infty} U^2 e^{iUy} f(U)\, dU;$$

$$\Theta^{(n)}(y) = i^n \int_{-\infty}^{+\infty} U^n e^{iUy} f(U)\, dU.$$

With y = 0 we have

$$[\Theta^{(n)}(y)]_{y=0} = i^n \int_{-\infty}^{+\infty} U^n f(U)\, dU. \tag{18}$$

The integral on the right side of (18) is called the nth moment of the random magnitude U. Let us note that

$$\int_{-\infty}^{+\infty} \varphi(U) f(U)\, dU$$

is the mathematical expectation or the average value of any function $\varphi(U)$. Therefore, the nth moment of U, equal to

$$a_n = \int_{-\infty}^{+\infty} U^n f(U)\, dU, \tag{19}$$

is nothing but the mathematical expectation or mean value of the nth power of the random magnitude U. In particular, a_1 is the mean value of U (in our case equal to zero), a_2 is the mean square value, etc.

Thus, (18) may be written in the form

$$[\Theta^{(m)}(y)]_{y=0} = i^m a_m, \tag{20}$$

and series (17) in the following manner:

$$\Theta(y) = 1 + \sum_{m=2}^{\infty} \frac{1}{m!} i^m a_m y^m. \tag{21}$$

The summation begins with m = 2, since we chose U to be centered at zero and $a_1 = 0$.

Taking the logarithm and expanding into series according to the formula $\ln(1+x) = x - \frac{1}{2}x^2 + \frac{1}{3}x^3 - \ldots$, we obtain a power series beginning with the second-order term. For the kth characteristic function we have

$$\ln \Theta_k(y) = \sum_{m=2}^{\infty} \frac{i^m a_{km}}{m!} y^m - \frac{1}{2}\left(\sum_{m=2}^{\infty} \frac{i^m a_{km}}{m!} y^m\right)^2 + \\ + \frac{1}{3}\left(\sum_{m=2}^{\infty} \frac{i^m a_{km}}{m!} y^m\right)^3 - \ldots, \tag{22}$$

i.e.,

$$\ln \Theta_k(y) = -\frac{a_{2k}}{2} y^2 + i\alpha_{3k} y^3 + \alpha_{4k} y^4 + i\alpha_{5k} y^5 + \ldots$$

$$\alpha_{3k} = -\frac{a_{3k}}{3!}; \qquad \alpha_{4k} = \frac{a_{4k}}{4!} - \frac{a_{2k}^2}{2(2!)^2}; \qquad \alpha_{5k} = \frac{a_{5k}}{5!} - \frac{a_{2k}a_{3k}}{2!3!};$$

$$\alpha_{6k} = -\frac{a_{6k}}{6!} + \frac{a_{3k}^2}{2(3!)^2} + \frac{a_{2k}a_{4k}}{2!4!} - \frac{a_{2k}^3}{3(2!)^3};$$

$$\alpha_{7k} = -\frac{a_{7k}}{7!} + \frac{a_{2k}a_{5k}}{2!5!} - \frac{a_{2k}^2 a_{3k}}{(2!)^2 3!} + \frac{a_{3k}a_{4k}}{3!4!};$$

$$\alpha_{8k} = \frac{a_{8k}}{8!} - \frac{a_{4k}^2}{2(4!)^2} - \frac{a_{2k}a_{6k}}{2!6!} - \frac{a_{3k}a_{5k}}{3!5!} + \frac{a_{2k}a_{3k}^2}{2!(3!)^2} + \\ + \frac{a_{2k}^2 a_{4k}}{(2!)^2 4!} - \frac{a_{2k}^4}{4(2!)^4};$$

$$\alpha_{9k}=\frac{a_{9k}}{9!}-\frac{a_{2k}a_{7k}}{2!7!}-\frac{a_{3k}a_{ck}}{3!6!}-\frac{a_{4k}a_{5k}}{4!5!}+\frac{a_{2k}^{2}a_{5k}}{(2!)^{2}\,5!}+$$

$$+\frac{2a_{2k}a_{3k}a_{4k}}{2!3!4!}+\frac{a_{3k}^{3}}{3\,(3!)^{3}}-\frac{a_{2k}^{3}a_{3k}}{(2!)^{3}\,3!};$$

$$\alpha_{10k}=-\frac{a_{10k}}{10!}+\frac{a_{5k}^{2}}{2\,(5!)^{2}}+\frac{a_{2k}a_{8k}}{2!8!}+\frac{a_{3k}a_{7k}}{3!7!}+\frac{a_{4k}a_{6k}}{4!6!}-\qquad(23)$$

$$-\frac{a_{2k}a_{4k}^{2}}{2!\,(4!)^{2}}-\frac{a_{2k}^{2}a_{6k}}{(2!)^{2}\,6!}-\frac{2a_{2k}a_{3k}a_{5k}}{2!3!5!}-\frac{a_{4k}a_{3k}^{2}}{4!\,(3!)^{2}}+\frac{3a_{2k}^{2}a^{2}_{k}}{2\,(2!)^{2}\,(3!)^{2}}+$$

$$+\frac{a_{2k}^{3}a_{4k}}{(2!)^{3}\,4!}-\frac{a_{2k}^{5}}{5\,(2!)^{5}}.$$

We shall write the logarithm of the characteristic function of the sum in the form

$$\ln\Theta(y)=-\frac{A}{2}y^{2}+iA_{3}y^{3}+A_{4}y^{4}+iA_{5}y^{5}+\dots,\qquad(24)$$

where

$$A=\sum_{k=1}^{M}a_{2k}\qquad(25)$$

and

$$A_{m}=\sum_{k=1}^{M}\alpha_{mk}.\qquad(26)$$

Taking the exponential of (24), we obtain

$$\Theta(y)=e^{-\frac{A}{2}y^{2}}e^{iA_{3}y^{3}+A_{4}y^{4}+\cdots};$$

$$\Theta(y)=e^{-\frac{A}{2}y^{2}}[1+(iA_{3}y^{3}+A_{4}y^{4}+iA_{5}y^{5}+\dots)+$$

$$+\frac{1}{2!}(iA_{3}y^{3}+A_{4}y^{4}+\dots)^{2}+\frac{1}{3!}(iA_{3}y^{3}+A_{4}y^{4}+\dots)^{3}+\dots].$$

Thus, according to (12)

$$P(y)=e^{-\frac{A}{2}y^2}\left[1+A_4y^4+\left(A_6-\frac{A_3^2}{2}\right)y^6+\left(A_8+\frac{A_4^2}{2}\right)y^8+\right.$$
$$\left.+\left(A_{10}-\frac{A_5^2}{2}+A_4A_6-\frac{A_3^2A_4}{2}-A_3A_7\right)y^{10}+\ldots;\right. \tag{27}$$

$$Q(y)=e^{-\frac{A}{2}y^2}\left[A_3y^3+A_5y^5+(A_7+A_3A_4)y^7+\right.$$
$$\left.+\left(A_9+A_4A_5+A_3A_6-\frac{1}{6}A_3^3\right)y^9+\ldots\right]. \tag{28}$$

Let us pass to the computation of the distribution function, according to (16).

The function sought is

$$f(U)=\frac{1}{\pi}\int_0^\infty e^{-\frac{A}{2}y^2}\left[1+\sum_{n=2}^{\infty}\beta_{2n}y^{2n}\right]\cos Uy\,dy+$$
$$+\frac{1}{\pi}\int_0^{+\infty}e^{-\frac{A}{2}y^2}\left[\sum_{n=1}^{\infty}\beta_{2n+1}y^{2n+1}\right]\sin Uy\,dy, \tag{29}$$

where

$$\beta_3=A_3,\ \beta_4=A_4,\ \beta_5=A_5,\ \beta_6=A_6-\frac{A_3^2}{2},$$
$$\beta_7=A_7+A_3A_4,\ \beta_8=A_8+\frac{A_4^2}{2},$$
$$\beta_9=\left(A_9+A_4A_5+A_3A_6-\frac{1}{6}A_3^3\right),$$
$$\beta_{10}=A_{10}-\frac{1}{2}A_5^2+A_4A_6-\frac{1}{2}A_3^2A_4-A_3A_7$$

and so forth.

The integrals entering into (29) are computed in the following manner:

$$\frac{1}{\pi}\int_0^{+\infty}e^{-\frac{A}{2}y^2}\cos Uy\,dy=\frac{1}{\sqrt{2\pi A}}e^{-\frac{U^2}{2A}}. \tag{30}$$

Let us denote (30) by $\psi(U, A)$. Then the integral of interest to us

$$\frac{1}{\pi}\int_0^{+\infty} e^{-\frac{A}{2}y^2} y^{2n} \cos Uy\, dy$$

may be computed by differentiating integral (30) with respect to the parameter A. We easily obtain

$$\frac{1}{\pi}\int_0^{+\infty} e^{-\frac{A}{2}y^2} y^{2n} \cos Uy\, dy = (-2)^n \psi_A^{(n)}. \tag{31}$$

Differentiating (30) with respect to U we obtain

$$\frac{1}{\pi}\int_0^{+\infty} e^{-\frac{A}{2}y^2} y \sin Uy\, dy = \frac{1}{\sqrt{2\pi A}} e^{-\frac{U^2}{2A}} \frac{U}{A}. \tag{32}$$

Let us denote (32) by $\Phi(U, A)$. Differentiating (32) with respect to the parameter A, we obtain a second series of interesting integrals:

$$\frac{1}{\pi}\int_0^{+\infty} e^{-\frac{A}{2}y^2} y^{2n+1} \sin Uy\, dy = (-2)^n \Phi_A^{(n)}. \tag{33}$$

The distribution function sought (29) appears as

$$f(U) = \frac{1}{\sqrt{2\pi A}} e^{-\frac{U^2}{2A}} \left[1 + \beta_4(-2)^2\frac{\psi^{(2)}}{\psi} + \beta_6(-2)^3\frac{\psi^{(3)}}{\psi} + \right.$$

$$\left. + \ldots + \beta_3(-2)^1\frac{\Phi^{(1)}}{\psi} + \beta_5(-2)^3\frac{\Phi^{(3)}}{\psi} + \ldots\right].$$

The expressions $\psi^{(n)}/\psi$ and $\Phi^{(n)}/\psi$ represent the polynomials U. We shall compute them by differentiating the right sides of (30) and (32) $\underline{n}$ times with respect to A. We shall write out the values of the first and the $\underline{n}$th polynomials for $\psi^{(n)}$:

$$
\left.\begin{aligned}
(-2)^2\frac{\psi^{(2)}}{\psi} &= \frac{3}{A^2}\left[1 - 2\left(\frac{U}{\sqrt{A}}\right)^2 + \frac{1}{3}\left(\frac{U}{\sqrt{A}}\right)^4\right];\\
(-2)^3\frac{\psi^{(3)}}{\psi} &= \frac{15}{A^3}\left[1 - 3\left(\frac{U}{\sqrt{A}}\right)^2 + \left(\frac{U}{\sqrt{A}}\right)^4 - \frac{1}{15}\left(\frac{U}{\sqrt{A}}\right)^6\right];\\
(-2)^4\frac{\psi^{(4)}}{\psi} &= \frac{105}{A^4}\left[1 - 4\left(\frac{U}{\sqrt{A}}\right)^2 + 2\left(\frac{U}{\sqrt{A}}\right)^4 - \frac{4}{15}\left(\frac{U}{\sqrt{A}}\right)^6 + \right.\\
&\qquad \left. + \frac{1}{105}\left(\frac{U}{\sqrt{A}}\right)^8\right];\\
(-2)^n\frac{\psi^{(n)}}{\psi} &= \frac{(2n-1)!!}{A^n}\sum_{k=0}^{n}\frac{(-1)^k}{(2k-1)!!}\binom{n}{k}\left(\frac{U}{\sqrt{A}}\right)^{2k}.
\end{aligned}\right\} \tag{34}
$$

In the same way for $\Phi^{(n)}$:

$$(-2)\frac{\Phi'}{\psi} = \frac{3}{A^2}\left[\frac{U}{\sqrt{A}} - \frac{1}{3}\left(\frac{U}{\sqrt{A}}\right)^3\right];$$

$$(-2)^2\frac{\Phi^{(2)}}{\psi} = \frac{15}{A^3}\left[\frac{U}{\sqrt{A}} - \frac{2}{3}\left(\frac{U}{\sqrt{A}}\right)^3 + \frac{1}{15}\left(\frac{U}{\sqrt{A}}\right)^5\right];$$

$$(-2)^3\frac{\Phi^{(3)}}{\psi} = \frac{105}{A^4}\left[\left(\frac{U}{\sqrt{A}}\right) - \frac{3}{3}\left(\frac{U}{\sqrt{A}}\right)^3 + \frac{1}{5}\left(\frac{U}{\sqrt{A}}\right)^5 - \frac{1}{105}\left(\frac{U}{\sqrt{A}}\right)^7\right].$$

$$(-2)^n\frac{\Phi^{(n)}}{\psi} = \frac{(2n+1)!!}{A^{n+1}}\sum_{k=0}^{n}\frac{(-1)^k}{(2k+1)!!}\binom{n}{k}\left(\frac{U}{\sqrt{A}}\right)^{2k+1}.$$

The distribution function sought is expressed finally in the following manner:

$$
\begin{aligned}
f(U) = \frac{1}{\sqrt{2\pi A}}e^{-\frac{U^2}{2A}}\Bigg[1 &+ \sum_{n=2}^{\infty}\beta_{2n}\frac{(2n-1)!!}{A^n}\sum_{k=0}^{n}\frac{(-1)^k}{(2k-1)!!}\binom{n}{k}\left(\frac{U}{\sqrt{A}}\right)^{2k} +\\
&+ \sum_{n=1}^{\infty}\beta_{2n+1}\frac{(2n+1)!!}{A^{n+1}}\sum_{k=0}^{n}\frac{(-1)^k}{(2k+1)!!}\binom{n}{k}\left(\frac{U}{\sqrt{A}}\right)^{2k+1}\Bigg].
\end{aligned} \tag{35}
$$

Here (n/k) is a binomial coefficient. The distribution function is expressed by computing the magnitudes U in terms of the mean square A, and also the higher moments, through which, according to (23), (26), and (30) it is possible to express the coefficients β_n.

Formula (35) was first applied to structure amplitudes by Karl and Hauptmann [19] and to structure products by us [20, 21].

3. Gaussian Representation of the Distribution Function of the Structure Amplitude for a Centrosymmetric Crystal

In this paragraph, by analyzing the general formula (35), we shall show that a Gaussian approximation to the distribution function satisfies completely the needs of structure analysis.

Let the sum of the random magnitudes w_k [formula (8)] be a unitary structure amplitude.

The mean value $\overline{w}_k$ of a component of the unitary structure amplitude for a crystal of any symmetry, and consequently also of $\overline{W}$, is equal to zero. Therefore U = W. The questions of the convergence of the series in (35) which interest us, and their values in comparison with unity will be clarified by using the example of a crystal of group P$\overline{1}$ with N identical atoms in the unit cell.

Then

$$U=\sum_{k=1}^{N/2}\frac{2}{N}\cos\alpha_k.$$

The moments of the addends of the magnitude U that occur in (25) and (26) were denoted by a_m. In our case all the addends are equal and the $\underline{m}$th moments of all $\frac{N}{2}$ addends are equal to one another. The values of the moments are as follows: $a_2=\frac{4}{N^2}\overline{\cos^2\alpha}$, $a_4=\frac{16}{N^4}\overline{\cos^4\alpha}$ etc. All the odd moments are equal to zero.

Taking into consideration that

$$\overline{\cos^{2n}\alpha}=\frac{1\cdot3\cdot5\ldots(2n-1)}{2\cdot4\cdot6\ldots2n}=\frac{(2n-1)\,!!}{2n\,!!},$$

we obtain

$$a_2 = \frac{2}{N^2}; \quad a_4 = \frac{6}{N^4}; \quad a_6 = \frac{20}{N^6}; \quad a_8 = \frac{70}{N^8}; \quad a_{10} = \frac{252}{N^{10}} \ldots$$

Hence the coefficients α_m obtained from (23) will take on the following values:

$$\alpha_4 = \frac{1}{4N^4} - \frac{1}{2N^4} = -\frac{1}{4N^4};$$

$$\alpha_6 = -\frac{1}{36N^6} + \frac{1}{4N^6} - \frac{1}{3N^6} = -\frac{1}{9N^6};$$

$$\alpha_8 = \frac{1}{576N^8} - \frac{1}{32N^8} - \frac{1}{36N^8} + \frac{1}{4N^8} - \frac{1}{4N^8} = \frac{11}{192N^8} \text{ etc.}$$

With the aid of (25) and (26) we obtain the moments of the sum, i.e., the moments of the structure amplitude. To do this, the last formulas must be multiplied by $\frac{N}{2}$ (since the moments of all the addends are equal). We obtain

$$A = \frac{1}{N}, \quad A_4 = -\frac{1}{8N^3}, \quad A_6 = -\frac{1}{18N^5}, \quad A_8 = -\frac{11}{384N^7} \text{ etc.}$$

Finally, with the aid of (29) we obtain the coefficients β

$$\beta_4 = \frac{1}{8N^3}, \quad \beta_6 = -\frac{1}{18N^5}, \quad \beta_8 = -\frac{11}{384N^7} + \frac{1}{128N^6} \text{ etc.}$$

Now (35), for the special case under study, will appear as

$$f(U) = \frac{1}{\sqrt{2\pi}} e^{-\frac{1}{2}x^2} \left\{1 - \frac{3}{8}\frac{1}{N}\left(1 - 2x + \frac{1}{3}x^2\right) - \right.$$
$$- \frac{5}{6}\frac{1}{N^2}\left(1 - 3x + x^2 - \frac{1}{15}x^3\right) + \left[\frac{105}{128}\frac{1}{N^2} - 3{,}01\frac{1}{N^3}\right]\left(1 - 4x + \right.$$
$$\left. + 2x^2 - \frac{4}{15}x^3 + \frac{1}{105}x^4\right) + \left[6{,}5\frac{1}{N^3} - 15\frac{1}{N^4}\right]\left(1 - 5x + \frac{10}{3}x^2 - \right.$$
$$\left.\left. - \frac{2}{3}x^3 + \frac{1}{21}x^4 - \frac{1}{945}x^5\right) + \ldots\right\},$$

where $x = (U\sqrt{N})^2$, the ratio of the square of the structure amplitude to its mean square.

Denoting by B_k, the polynomials of the kth order, which occur in curly brackets, we obtain for N = 10:

$$f(x)=\frac{1}{\sqrt{2\pi}}e^{-\frac{1}{2}x^2}\{1-0.03B_2-0.01B_3+0.006B_4+0.005B_5\ldots\}$$

and for N = 40:

$$f(x)=\frac{1}{\sqrt{2\pi}}e^{-\frac{1}{2}x^2}\{1-0.01B_2-0.0005B_3+0.0006B_4+0.0001B_5\ldots\}.$$

The values of B_k for two values of x are as follows:

x	B_2	B_3	B_4	B_5
1	$-2/3$	$-16/15$	$-19/15$	$-4/3$
5	$-2/3$	$+8/3$	$+4$	$+3$

The convergence of the series is determined by the polynomial B_k, i.e., by the value of $\hat{F}$ for which we wish to compute the probability, and by the number of atoms in the cell N.

From Fig. 8, which shows the course of the curves of B for values of the interval of the magnitude x which might have meaning, we see that up to values of x of the order of seven to eight (and we are certain not to meet with higher values) the magnitudes of B_k are smaller than six.

The numerical coefficients of the powers of 1/N increase during the transition from one term of the series to the next. Hence, it follows that the series will converge if the values of B and of the powers of 1/N decrease more rapidly than the coefficients increase. It is clear that the greater the number of atoms N and the smaller x, the better the series converges.

If $\hat{F}$ approaches unity, then the series will diverge for any N. It must be so, because unity is the terminal value of $\hat{F}$.

On the other hand, it is possible to assign such a value of x to each N that the series will begin to diverge. However, the only thing that is important is whether the series diverges at magnitudes of x which we are likely to meet experimentally. Evidently, one does not meet with magnitudes exceeding the average by more than five or six times. The value x = 5 already converts the exponential factor into 10^{-6}. Consequently, we only have to discover for which N the series (35) will converge for values of x smaller than that indicated. Of course, for the sake of rigor, one ought to compute the higher terms of the series as well. But the computation is cumbersome, and

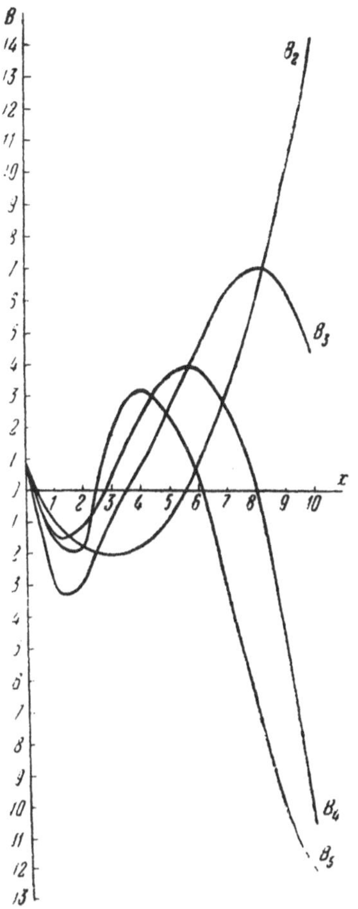

Fig. 8. The polynomials B (x). $B_2 = 1 - 2x + \frac{1}{3}x^2$; $B_3 = 1 - 3x + x^2 - \frac{1}{15}x^3$; $B_4 = 1 - 4x + 2x^2 - \frac{4}{15}x^3 + \frac{1}{105}x^4$; $B_5 = 1 - 5x + \frac{10}{3}x^2 - \frac{2}{3}x^3 + \frac{1}{21}x^4 - \frac{1}{945}x^5$.

the result is clear from the numerical example. There are, generally speaking, three regions of values for $U/\sqrt{A}$: the region of the Gaussian approximation, of the converging series, and of the diverging series.

Careful analysis assures us that as soon as the series begins to converge, the Gaussian approximation is fulfilled with good accuracy. Thus, there is no need to use the series for computing, and the conclusion given in (35) must be considered as a proof of the applicability of the Gaussian approximation for even a slightly complex crystal.

It is rather hopeless to evaluate the deviations from the Gaussian distribution with the aid of (35) – the series converges slowly, and in order to evaluate these deviations from the Gaussian distribution it is necessary to take an impossibly large number of terms.

4. The Dependence of the Amplitude Distribution Function on the Structure

Thus, for a crystal with more than ten to twenty atoms, the distribution function of the structure amplitude of a crystal possessing an inversion center may be written

$$f(F) = \frac{1}{\sqrt{2\pi A}} e^{-\frac{F^2}{2A}}, \qquad (36)$$

where A is the second order moment of the unitary structure amplitude computed according to (25).

This distribution function is determined uniquely by the magnitude A.

Therefore, some general theorems pertaining to this magnitude are essential. The basic results stated in this and the following paragraphs were first derived by Wilson [22].

Since, by definition, A is dependent on the form of the structure amplitude formula, the value A will be different both for one type of reflections in different translation groups, and for different types in a single translation group.

To compute the magnitudes of the distribution parameter A (in Wilson's work and in that of his co-workers this magnitude is symbolized by S) it is necessary to square the unitary structure amplitudes and to take their average.

If the number of nodes per cell is denoted by $\underline{z}$, and by ϵ, the multiplicity of the symmetry element of reciprocal space on which lie the reciprocal lattice points of the group of reflections under consideration, then

$$A = z\varepsilon \sum_{1}^{N} n_j^2. \tag{37}$$

Let us illustrate (37) by examples.

Class C_{2h}, Group P, $z=1$.

Nodes hkl, $0kl$ & $hk0$, $\varepsilon=1$, $A=16\sum_{1}^{N/4} n_j^2\,\overline{\cos^2\alpha}\,\overline{\cos^2\beta}=\sum_{1}^{N} n_j^2.$

Nodes $h0l$, $\varepsilon=2$, $A=16\sum_{1}^{N/4} n_j^2\,\overline{\cos^2\alpha}=2\sum_{1}^{N} n_j^2.$

Nodes $0k0$, $\varepsilon=2$, $A=16\sum_{1}^{N/4} n_j^2\,\overline{\cos^2\beta}=2\sum_{1}^{N} n_j^2.$

Class C_{2h}, Group C, $z=2$.

Nodes hkl, $hk0$ & $0kl$, $\varepsilon=1$, $A=64\sum_{1}^{N/8} n_j^2\,\overline{\cos^2\alpha}\,\overline{\cos^2\beta}=2\sum_{1}^{N} n_j^2.$

Nodes $h0l$, $\varepsilon=2$, $A=64\sum_{1}^{N/8} n_j^2\,\overline{\cos^2\alpha}=4\sum_{1}^{N} n_j^2.$

Nodes $0k0$, $\varepsilon=2$, $A=64\sum_{1}^{N/8} n_j^2\,\overline{\cos^2\beta}=4\sum_{1}^{N} n_j^2.$

Class D_{2h}, Group F, $z=4$.

Nodes hkl, $\varepsilon=1$, $A=32^2\sum_{1}^{N/32} n_j^2\,\overline{\cos^2\alpha}\,\overline{\cos^2\beta}\,\overline{\cos^2\gamma}=4\sum_{1}^{N} n_j^2.$

Nodes $h0l$, $\varepsilon=2$, $A=32^2\sum_{1}^{N/32} n_j^2\,\overline{\cos^2\alpha}\,\overline{\cos^2\gamma}=8\sum_{1}^{N} n_j^2.$

Nodes $h00$, $z = 4$, $A = 32^2 \sum_1^{N/32} n_j^2 \overline{\cos^2 \alpha} = 16 \sum_1^N n_j^2$.

Class D_{2h}. Group P, $z = 1$.

Nodes hkl, $z = 1$, $A = 64 \sum_1^{N/8} n_j^2 \overline{\cos^2 \alpha}\, \overline{\cos^2 \beta}\, \overline{\cos^2 \gamma} = \sum_1^N n_j^2$.

Nodes $hk0$, $0kl$, $h0l$, $z = 2$, $A = 64 \sum_1^{N/8} n_j^2 \overline{\cos^2 \alpha}\, \overline{\cos^2 \beta} = 2 \sum_1^N n_j^2$.

Nodes $00l$, $0k0$, $h0l$, $z = 4$, $A = 64 \sum_1^{N/8} n_j^2 \overline{\cos^2 \alpha} = 4 \sum_1^N n_j^2$.

Strictly speaking one should consider the averaging of the different classes (in the sense of page 69) of structure amplitudes. However, it is evident that the formulas for A will be identical for all these classes, and will be affected only by the presence of zeros in the indices hkl. This follows from the fact that the average value of $\overline{\sin^2 \alpha}$ is equal to the average value of $\overline{\cos^2 \alpha}$.

Let us denote by $\hat{\bar{F}}^2$ (in Wilson's notation <I>) the mean value of the structure factor, taken from the experimental data for all the nodes of the reciprocal lattice of similar multiplicity. The constant A is computed as the mean square of the expression for the structure amplitude, formed for any group of nonzero reflections. Therefore,

$$\left.\begin{array}{ll} \overline{\hat{F}^2} = A & \text{in the absence of extinctions,} \\ \overline{\hat{F}^2} = \frac{1}{m} A & \text{when systematic extinctions are present, and} \\ \left(1 - \frac{1}{m}\right) & \text{is the fraction of reflections for those axes or planes in which extinctions are observed.} \end{array}\right\} \quad (38)$$

Comparing (38) and (37), we see that for indices of the general type, independent of the centering of the cell and of the symmetry,

$$\overline{\hat{F}^2} = \sum_1^N n_j^2. \quad (39)$$

If we are sure of the presence of some kind or other of systematic extinctions, then A acquires the meaning of the average $\hat{\bar{F}}^2$, taken over the reflections that are present.

However, uncertainty as to the reality of any symmetry element that creates extinctions is not removed by computations of $\hat{\bar{F}}^2$.

For example, let there be a pseudoglide plane in a crystal with a monoclinic cell of symmetry $\bar{1}$. Let the coordinates of the atoms n_j be xyz; $\overline{xyz}$; x, a−y, z + $\frac{1}{2}$; x, a + y, $\frac{1}{2}$ − z. The structure amplitude formula can then written in the form

$$\hat{F}=4\sum n_j \cos 2\pi\left(hx+lz+\frac{l}{2}+\frac{ka}{2}\right)\cos 2\pi\left(ky-\frac{l}{2}-\frac{ka}{2}\right).$$

For the nodes h0*l* this expression does not differ in any way from the one in which a glide plane is present. In this manner the values of A and m, and consequently of $\hat{\bar{F}}^2$ also will be the same.

This should be considered rather as a positive fact, since such insensitivity to the real symmetry makes it possible to find $\sum n_j^2$ correctly from experiment, and this is essential (see page 101). Thus, for all centrosymmetric crystals (37) is always satisfied. Only inexact centrosymmetry of the crystal, as we shall see in the following paragraph, can break (37).

Success, or at any rate, ease in carrying out x-ray structure analysis is dependent basically on the fraction of "strong" structure amplitudes present in the experimental material. Let us write out the following characteristic data from a table of values of the normalized Gaussian function.

$\lvert\hat{F}\rvert >$	$\sqrt{A}$	$2\sqrt{A}$	$3\sqrt{A}$	$4\sqrt{A}$
The fraction of the structure amplitudes	0.3174	0.0456	0,0027	Less than 0.0001

Thus, experimentally, structure amplitudes greater than four times the root mean square are almost impossible. Taking $\hat{\bar{F}}^2 = 16\ A^2$ as the extreme upper limit of the structure factor, we obtain, for the case of similar atoms $\left(n_j = \frac{1}{N}\right)$:

$$\hat{F}_{\max}=4\sqrt{\frac{z\varepsilon}{N}}.$$

This formula also indicates the region in which statistical considerations apply. Since $\hat{F}^2_{max} < 1$ this formula cannot be applied to the case of a small number of atoms in general positions.

For the group C^5_{2h} for example, we have

N	16	20	30	40	50	60	70	80	90	100	120	140	160	180	200
Nodes hkl $\hat{F}_{max}$	1	0.89	0.73	0.63	0.56	0.52	0.48	0.45	0.42	0.40	0.36	0.34	0.31	0.30	0.28
Nodes $h0l$ and $0k0$ $\hat{F}_{max}$	—	—	—	0.89	0.79	0.73	0.68	0.63	0.59	0.56	0.51	0.48	0.43	0.42	0.40

The presence of a heavy atom decreases N_{eff}, the "effective" number of atoms $\left\{\sum n_j^2\right\}^{-\frac{1}{2}}$; the fraction of strong structure amplitudes increases, and the carrying out of the structure analysis by any method is facilitated.

5. Deviation of a Crystal Center from Centrosymmetry. "Complete" Loss of the Inversion Center

A statistically important feature of all centrosymmetric crystals is the fact that deviation from or approximation to any element of symmetry (while retaining $\bar{1}$) does not bring about a change of the distribution parameter A, which, for the reflections hkl of any centrosymmetric crystal, is equal to Σn_j^2.

What is the situation with regard to deviation of the crystal from the symmetry $\bar{1}$? This deviation brings with it the imaginary part of the structure amplitude. The values F_{hkl} must now be distributed not along a line, but in the complex plane. In other words, the distribution function of $F = A + Bi$ is affected not by one, but by two variables – the real part A and the imaginary part B of the structure amplitude.

The distribution function $f(A, B)$ may differ in appearance depending on the degree of deviation of the crystal from the symmetry $\bar{1}$.

For the real and imaginary parts of F we shall easily prove the Gaussian distribution by the methods that have just been studied. These functions will be sections of $f(A, B)$, parallel to the axis of A or the axis of B. However, for different A (or B) the distribution functions B (or A) may differ depending on the dispersion. Therefore, in the general case, f(A, B) will have symmetry only with respect to the axes of coordinates.

The smaller the deviation from the symmetry $\bar{1}$, the more the distribution function will contract toward the real axis.[2] On the other hand, the limiting deviation from the symmetry $\bar{1}$ will arise for a circular symmetry of f(A, B). Such an aspect of f(A, B) may be considered the definition of the ideal noncentrosymmetric case.[3]

Thus, a complete loss of symmetry $\bar{1}$ occurs if the distribution functions of the real and imaginary parts of $\hat{F}$ are similar, i.e., $[f(A)]_{B=C} = [f(B)]_{A=C}$ where C is any number. It is not hard to see that such a distribution function f (A, B) exists when the contributions to the structure amplitude F = A + Bi by different atoms may be considered as random independent magnitudes.

Only when there is complete loss of the symmetry $\bar{1}$ are the dispersions for $A = \sum_1^N n_j \cos \alpha_j$ and $B = \sum_1^N n_j \sin \alpha_j$ identical and equal to

$$\overline{A^2} = \overline{B^2} = \frac{1}{2} \sum_1^N n_j^2. \tag{40}$$

Then the distribution function appears as

$$f(A, B) = f(A) f(B) = \frac{1}{\pi \sum n_j^2} e^{-\frac{A^2+B^2}{\sum n_j^2}}. \tag{41}$$

[2]Since by rotating the axes of coordinates by 90°, it is always possible to transform the real part into the imaginary one and vice versa, the case of purely imaginary F's has no place. Depending on the choice of origin of the cell, the function f(A, B) turns in the complex plane. Probably, it is possible on this basis to give a method for choosing uniquely the origin of coordinates in a crystal without an inversion center. [This argument is not correct—Translator.]
[3]A shift of origin away from an inversion center by Δx, Δy, Δz multiplies each structure factor by exp 2πi (hΔx + kΔy + lΔz). Then, a shift in origin away from either a true or approximate center can always produce circular symmetry in f(A,B). The author's argument is based on his overlooking this fact — Translator.

Let us consider a partial loss of the inversion center[4] by two examples. Let there be N atoms in the cell of which M may be divided into $\frac{M}{2}$ pairs, connected by a center of symmetry, and the remaining N − M be randomly distributed. Then

$$\hat{F} = 2\sum_{j=1}^{M/2} n_j \cos\alpha_j + \sum_{j=M+1}^{N} n_j \cos\alpha_j + i\sum_{j=M+1}^{N} n_j \sin\alpha_j. \quad (42)$$

With this notation we assume all the α's to be random, independent quantities. Then

$$\overline{A^2} = \sum_{j=1}^{M} n_j^2 + \frac{1}{2}\sum_{j=M+1}^{N} n_j^2; \qquad \overline{B^2} = \frac{1}{2}\sum_{j=M+1}^{N} n_j^2. \quad (43)$$

The distribution function f(A, B) = f(A) f(B) has the form

$$f(A, B) = \frac{1}{\pi\sqrt{\overline{A^2}\,\overline{B^2}}}\, e^{-\frac{A^2}{2\overline{A^2}} - \frac{B^2}{2\overline{B^2}}}. \quad (44)$$

When M = 0 the formula is converted into (41), and, when M = N, into (36).

As a second example, let us consider the case when pairs of atoms are shifted by small magnitudes Δ_j from positions corresponding to an inversion center. Dividing the atoms into pairs, and confining ourselves to the first approximation ($\cos 2\pi \mathbf{H}\Delta_j = 1$, $\sin 2\pi \mathbf{H}\Delta_j = 2\pi \mathbf{H}\Delta_j$), it is always possible to write the expression for the structure amplitude in the form

$$\hat{F} = 2\sum_{1}^{N/2} n_j \cos\alpha_j + 2i\sum_{1}^{N/2} n_j (2\pi\mathbf{H}\Delta_j)\cos\alpha_j.$$

Evidently, the distribution (44) will appear also in this case with the sole difference that $\overline{A^2}$ and $\overline{B^2}$ depend on the length of the reciprocal vector. This means that for small **H** the structure may be centrosymmetric to the accuracy of the experiment, while for large **H** a deviation from this symmetry is observed.

[4]Such a possibility was discussed in the work of A. I. Kitaigorodskii [23] and A. Wilson [24].

6. Statistical Differences Between Centrosymmetric Crystals and Crystals without Inversion Centers

Although the mean value of the structure factor $\overline{\hat{F}^2}$ does not change whether an inversion center is present or not, and, for nodes with general indices is always [see (40) and (43)] $\overline{\hat{F}^2} = \sum n_j^2$, nevertheless there are essential statistical differences between crystals with and without the symmetry $\overline{1}$.

The basis for this difference lies in the different form of the distribution function of the structure factor $|\hat{F}|^2$.

It is to be noted that when the probability density is f(x), then f(x)dx is the probability that x lies in the interval from x to x + dx. Furthermore, if what is required is the probability density not of x, but of some function of x, let us say φ (x), then φ is found from the condition

$$f[\varphi(x)]\, d[\varphi(x)] = f(x)\, dx.$$

If, for example, the probability of x^2 is sought, then

$$f_1(x^2)\, 2x\, dx = f(x)\, dx,$$

or

$$2f_1(x^2)\,|x|\, d\,|x| = f(x)\, dx.$$

Applying this result to a centrosymmetric crystal, we obtain, as a consequence of (36):

$$f(\hat{F}^2) = \frac{1}{\sqrt{2\pi\Sigma n_j^2}\,|\hat{F}|}\, e^{-\frac{\hat{F}^2}{2\Sigma n_j^2}}. \tag{45}$$

For a crystal without a symmetry center we proceed from (41). The probability that $\hat{F}^2$ lies between $\hat{F}^2$ and $\hat{F}^2 + d(\hat{F}^2)$, is equal to the probability that the point (A, B) lies within a ring with the radii $\sqrt{\hat{F}^2}$ and $\sqrt{\hat{F}^2} + d(\sqrt{\hat{F}^2})$. Hence

$$f(\hat{F}^2)\, d(\hat{F}^2) = 2\pi f(A,\ B) \sqrt{\hat{F}^2}\, d\left(\sqrt{\hat{F}^2}\right) =$$

$$= \frac{1}{\sum n_j^2}\, e^{-\frac{\hat{F}^2}{\Sigma n_j^2}}\, d(\hat{F}^2). \tag{46}$$

Since (45) and (46) differ, the possibility arises that a purely x-ray-structure experiment can enable one to detect the absence of an inversion center.

In practice this may be done for example, by the following methods: 1) by computing integrated distribution functions, 2) by computing the ratios $|\hat{F}|^2/\bar{\hat{F}}^2$, 3) by computing the mean square deviation of $\hat{F}^2$.

All these methods have been suggested and worked out by Wilson and his co-workers [25-27].

Let us compute the integrated distribution functions $N(\zeta)$ where $\zeta = \hat{F}^2/\bar{\hat{F}}^2$. The number $N(\zeta)$ is the fraction of the reflections the structure factors of which are less than or equal to ζ.

Rewriting (45) in the form

$$f(\zeta)\, d\zeta = (2\pi\zeta)^{-\frac{1}{2}} e^{-\frac{1}{2}\zeta} d\zeta$$

and integrating from 0 to ζ, we obtain

$$N(\zeta) = \int_0^{\zeta} f(\zeta)\, d\zeta = \frac{2}{\sqrt{\pi}} \int_0^{\left(\frac{1}{2}\zeta\right)^{\frac{1}{2}}} e^{-t^2} dt = \operatorname{erf} \sqrt{\frac{\hat{F}^2}{2\bar{\hat{F}}^2}}. \tag{47}$$

This function is tabulated; it is nothing but the so-called probability integral:

$_1N(\zeta)$	0,2	0,4	0.6	0.8	0,98	0.9999
ζ	0.06	0.27	0.72	1.65	5.4	15.1
$\lvert\hat{F}\rvert/\sqrt{\bar{\hat{F}}^2}$	0.25	0.52	0.85	1.28	2.3	3.9

Rewriting (46) in the form

$$f(\zeta)\, d\zeta = e^{-\zeta} d\zeta$$

and integrating from 0 to ζ, we obtain

$$N(\zeta) = 1 - e^{-\zeta}. \tag{48}$$

A table of values of (48) has the form

$_1N(\zeta)$	0.2	0.4	0.6	0.8	0.98	0.9999
ζ	0.22	0.51	0.92	1.60	3.9	9.2
$\lvert\hat{F}\rvert/\sqrt{\bar{\hat{F}}^2}$	0.47	0.71	0.96	1.28	1.98	3.03

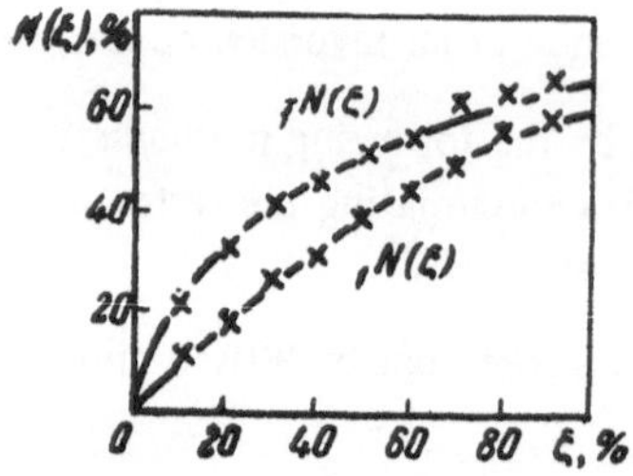

Fig. 9. Graphic comparison of the functions ${}_1N(\zeta)$ and ${}_{\bar{1}}N(\zeta)$. Experimental points are marked for ephedrinehydrochloride; projections [001] and [010].

Comparison of these two tables shows that the course of the curves differs greatly. In a crystal that has completely lost an inversion center, the fraction of weak reflections is considerably smaller: for example, a crystal without an inversion center has about 20% of its amplitudes less than half the root mean square, but for a crystal with an inversion center this number is about 40%.

At large values of ζ the curves intersect, and ${}_1N(\zeta)$ goes higher than ${}_{\bar{1}}N(\zeta)$. It is clear that this must be so, since $\overline{\hat{F}^2}$ is identical for crystals with and without an inversion center. The share of strong reflections is greater in a crystal without an inversion center.

The course of the curves ${}_1N(\zeta)$ and ${}_{\bar{1}}N(\zeta)$ is shown on Fig. 9. The experimental data are marked by crosses for the indices of two zones (one centrosymmetric, and the other not) of l-ephedrinehydrochloride ([28] sec. 1).

The second method of analysis of centrosymmetricity consists in determining the ratios

$$\rho \equiv \frac{\overline{|\hat{F}|}^2}{\overline{|\hat{F}|^2}}. \tag{49}$$

The computation of ρ for a crystal without an inversion center gives, with the aid of (46):

$$\overline{|\hat{F}|} = \frac{1}{\sum n_j^2} \int_0^{\infty} \hat{F} e^{-\frac{\hat{F}^2}{\sum n_j}} \, d\hat{F} = \frac{1}{2}\left(\pi \sum n_j^2\right)^{\frac{1}{2}}. \tag{50}$$

For crystals with a center of symmetry we have, proceeding from (45)

$$\overline{|\hat{F}|} = \frac{1}{\sqrt{2\pi \sum n_j^2}} \int_0^{\infty} e^{-\frac{\hat{F}^2}{\sum n_j^2}} \, d\hat{F} = \left(\frac{2}{\pi} \sum n_j^2\right)^{\frac{1}{2}}. \tag{51}$$

$$\left.\begin{array}{ll}\text{For crystals with an inversion center} & \rho=\frac{2}{\pi}=0.637; \\ \text{and} & \\ \text{for crystals without an inversion center} & \rho=\frac{\pi}{4}=0.785.\end{array}\right\} \tag{52}$$

Let us examine the third criterion of centrosymmetricity; the magnitude of the specific mean square deviation

$$v\equiv\overline{\left(\frac{\hat{F}^2-\overline{\hat{F}^2}}{\overline{\hat{F}^2}}\right)^2}. \tag{53}$$

Let us write the square of the structure amplitude,

$$\hat{F}^2=\sum_j\sum_k n_j n_k \exp i(\alpha_j-\alpha_k)$$

in the form

$$\hat{F}^2=\sum n_j^2+\sum_{j\neq k}\sum n_j n_k \exp i(\alpha_j-\alpha_k).$$

We have furthermore

$$\begin{aligned}(\hat{F}^2-\sum n_j^2)^2&=\sum_{\substack{j\neq k\\ i\neq l}} n_j n_k n_i n_l \exp i(\alpha_i+\alpha_j-\alpha_k-\alpha_l)=\\ &=\sum_{j\neq k} n_j^2 n_k^2+\sum_{i\neq j\neq k\neq l} n_j n_k n_i n_l \exp i(\alpha_i+\alpha_j-\alpha_k-\alpha_l).\end{aligned} \tag{54}$$

When averaged, the fourfold sum turns out to be zero. Observing that

$$\sum_{j\neq k} n_j^2 n_k^2=(\sum n_j^2)^2-\sum n_j^4, \tag{55}$$

and recalling (39), we obtain

$$v=1-\frac{\sum n_j^4}{(\sum n_j^2)^2} \tag{56}$$

for a crystal without an inversion center.

Let us now carry out an analogous computation for a centrosymmetric crystal.

Proceeding from

$$\hat{F}^2 = 4 \sum_{j}^{\frac{1}{2}N} \sum_{k}^{\frac{1}{2}N} n_j n_k \cos\alpha_j \cos\alpha_k,$$

separating out the terms with j = k and using the double angle formula we obtain

$$\hat{F}^2 = \sum_{1}^{N} n_j^2 + 2\sum_{1}^{N/2} n_j^2 \cos 2\alpha_j + 4\sum_{j\neq k}^{N/2}\sum^{N/2} n_j n_k \cos\alpha_j \cos\alpha_k.$$

Hence

$$\begin{aligned}(\hat{F}^2 - \overline{F^2})^2 = {} & 4\sum_{j}^{N/2}\sum_{k}^{N/2} n_j^2 n_k^2 \cos 2\alpha_j \cos 2\alpha_k + \\ & + 8\sum_{j}^{N/2}\sum_{k\neq l}^{N/2}\sum^{N/2} n_j^2 n_k n_l \cos 2\alpha_j \cos\alpha_k \cos\alpha_l + \\ & + 16\sum_{j\neq k}^{N/2}\sum \sum_{l\neq i}^{N/2}\sum n_j n_k n_l n_i \cos\alpha_j \cos\alpha_k \cos\alpha_l \cos\alpha_i.\end{aligned} \quad (57)$$

When averaged, only those terms of the sum are preserved into which the cosines enter in even powers:

$$\overline{(F^2 - \hat{F}^2)^2} = 4\sum_{j}^{N/2} n_j^4 \overline{\cos^2 2\alpha_j} + 32\sum_{j}^{N/2}\sum_{k} n_j^2 n_k^2 \overline{\cos^2\alpha_j \cos^2\alpha_k}.$$

The number of terms unequal to zero is doubled in the double sum, as opposed to the analogous sum for crystals without an inversion center, since for each pair of indices j and k there are four addends of the type $n_j^2 n_k^2 \cos^2\alpha_j \cos^2\alpha_k$, while the double summation takes into account only two addends. Using $\overline{\cos^2\alpha} = \frac{1}{2}$, and taking into account (55) we obtain

$$v = 2 - 3\frac{\sum n_j^4}{\left(\sum n_j\right)^2} \quad (58)$$

for a crystal with an inversion center.

It is useful to write the formulas for the mean square deviation v for identical atoms in the cell:

$$\left.\begin{aligned} v &= 1 - \frac{1}{N} \quad \text{for a crystal without } \bar{1}, \\ v &= 2 - \frac{3}{N} \quad \text{for a crystal with } \bar{1}. \end{aligned}\right\} \tag{59}$$

For sample computations the second term in the expression for v may be neglected.

Such computations were carried out for a series of crystals with the following rather good results:

Material	Space Group	v_{exp}	$\sim v_{theor}$
Tetraphenylcyclobutane zone $h0l$	$P2_1/a$	2.13	2
Nitrosyl Perchlorate zone hkl	Cc	1.25	1
Menthol, zone $hk0$	$C3_1$	1.05	1
Nitrogen Tetroxide zone $0kl$	Im3	2.28	2

7. The Distribution Functions of Structure Amplitudes and Hypersymmetry

A very important result of the preceding paragraphs is the relationship

$$\overline{F^2} = \sum n_j^2, \tag{60}$$

fulfilled with good accuracy for reflections with general indices for crystals of any symmetry. This important equation is insensitive also to any supersymmetry (or hypersymmetry) i.e., to any ordered arrangement of atoms within the cell. That is to say, such additional "noncrystallographic" symmetry elements of parts of the cell may lead to a weakening of some and to a strengthening of other reflections, but the mean square of the structure amplitude will remain the same. One has only to keep in mind that averaging in reciprocal space should be carried out over spherical regions or rings, and not over rows or planes of the reciprocal lattice. Only in the first case will $\alpha_j = 2\pi Hr_j$ be really uniformly distributed over the trigonometric circle and (60) arise.

We saw in the last paragraph that $\overline{F^2}$ is identical for crystals with and without an inversion center, but can be distinguished by their intensity distributions. We showed in the preceding paragraph that the intensity distribution for a crystal with an inversion center and for one completely without it differ greatly. However, cases of a partial loss of an inversion center are possible. Thus, intensity distribution curves lying between the curves in Fig. 9 shown above are possible.

We shall show in this paragraph that, in the case of supersymmetry, distributions lying above (for small $|\hat{F}|$) the distribution curves are possible for a centrosymmetric crystal.

Let us first of all analyze a particular case of hypersymmetry of the kind where, in addition to the crystallographic center of symmetry, the cell has two more "centers of symmetry" each of which affects only half of the cell content. Crystallographers have come across such cases several times.

Let us denote by $\pm \mathfrak{r}$ the coordinates of these hypercenters of inversion. Each atom is multiplied by hypercenters into four atoms with the coordinates $\pm(\mathbf{r}_j + \mathfrak{r})$ and $\pm(\mathbf{r}_j - \mathfrak{r})$, where $\mathbf{r}_j$ is the coordinate of the atom with respect to the hypercenter. The contribution of these four atoms to the structure amplitude, is equal to

$$\begin{aligned} n_j \{ \exp 2\pi i (\mathbf{r}_j + \mathfrak{r}) \mathbf{H} + \exp - 2\pi i (\mathbf{r}_j + \mathfrak{r}) \mathbf{H} + \\ \exp 2\pi i (\mathbf{r}_j - \mathfrak{r}) \mathbf{H} + \exp - 2\pi i (\mathbf{r}_j - \mathbf{r}) \mathbf{H} \} = \\ = 4 n_j \cos 2\pi \mathbf{r}_j \mathbf{H} \cos 2\pi \mathfrak{r} \mathbf{H}. \end{aligned} \tag{61}$$

It is useful to note that hypersymmetry leads to a redistribution of the intensity. It is clear that, in the planes of reciprocal space which satisfy the condition $\cos 2\pi \mathfrak{r} \mathbf{H} = 1$, there will be an increased intensity, and in the planes where $\cos 2\pi \mathfrak{r} \mathbf{H} = 0$ the weights of the nodes will be zero. Hence, the truth of what has been said in the beginning of this paragraph is evident as concerns the necessity of averaging $\hat{F}^2$ in (60) over spheres or rings in reciprocal space.

Viewing $\mathbf{r}_j \mathbf{H}$ as a random magnitude, we may consider that for each constant value of $4\mathfrak{r}\mathbf{H} = u$ (the coefficient 4 is included for convenience) the distribution of the structure amplitude

$$\hat{F} = 4 \sum_{j=1}^{N/4} n_j \cos 2\pi \mathbf{r}_j \mathbf{H} \cos 2\pi \mathfrak{r} \mathbf{H}$$

will be Gaussian, with the dispersion

$$2 \cos^2 2\pi \mathfrak{r} \mathbf{H} \sum n_j^2 = D.$$

It is as though the distribution were of those $\hat{F}$'s, the nodes of which fall on the plane $4\mathfrak{r}\mathbf{H} = u$.

The distribution function sought may be considered as the sum (integral) of similar distribution functions:

$$P(\hat{F})\,d\hat{F}=\frac{1}{2}\int\limits_{-1}^{+1}\frac{1}{\sqrt{2\pi D}}e^{-\frac{\hat{F}^2}{2D}}d\hat{F}\,du.$$

Integration, as a matter of fact, is carried out over rll from 0 to 90°. Introducing the abbreviation $t = \tan\frac{1}{2}\pi u$ we obtain

$$P(\hat{F})\,d\hat{F}=2\pi^{-1}\left(4\pi\sum n_j^2\right)^{-\frac{1}{2}}\int\limits_0^{+\infty}\exp\left[-\hat{F}^2(1+t^2)/4\sum n_j^2\right]\times$$

$$\times(1+t^2)^{-\frac{1}{2}}d\hat{F}dt.$$

or, going over to the distribution of structure factors,

$$P(\hat{F}^2)\,d(\hat{F}^2)=\pi^{-1}\left(4\pi\sum n_j^2\right)^{-\frac{1}{2}}\hat{F}|^{-1}\int\limits_0^{+\infty}\exp\left[-\hat{F}^2(1+t^2)/4\sum n_j^2\right]\times$$

$$\times(1+t^2)\,d(\hat{F}^2)\,dt$$

We see that the mean value of the structure factor is derived as before from (60), i.e., we emphasize once more that (60) preserves its validity in the presence of hypersymmetry. Denoting $\frac{\hat{F}^2}{\bar{\hat{F}}^2}=\zeta$, we obtain for the fraction of the structure factors smaller than ζ:

$$N(\zeta)=\pi^{-1}(4\pi)^{-\frac{1}{2}}\int\limits_0^{\zeta}\int\limits_0^{+\infty}\zeta^{-\frac{1}{2}}\exp\left[\frac{1}{4}\zeta(1+t^2)\right](1+t^2)\,dt\,d\zeta. \quad (62)$$

The appearance of (62) for small ζ (smaller than unity) is shown on Fig. 10. The curve lies above the distribution curve for crystals with an inversion center.

Formulas for other supersymmetric structures may also be derived by an analogous method.

In Rodgers and Wilson's [27] work, a considerable number of cases of hypersymmetry, such as might be met in practice, were analyzed.

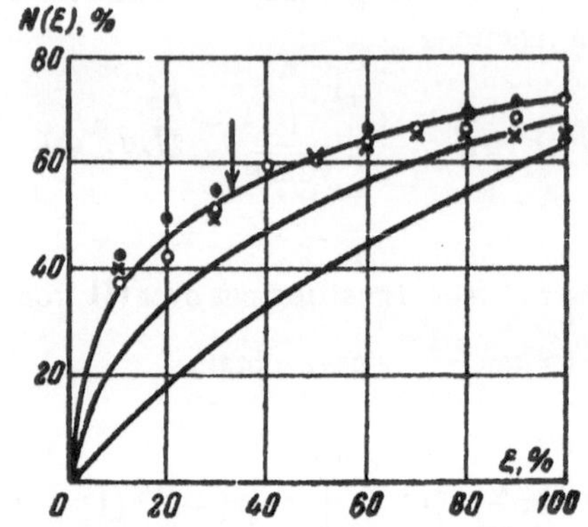

Fig. 10. Case of supersymmetry. The intensity distribution curves for pyrene (×) 1:1:6:6-tetraphenyl hexapentene (○), flavanone (●). Theoretical curves: the lower is for the acentric case, the middle is for the centrosymmetric case, the upper is for the hypercentric case.

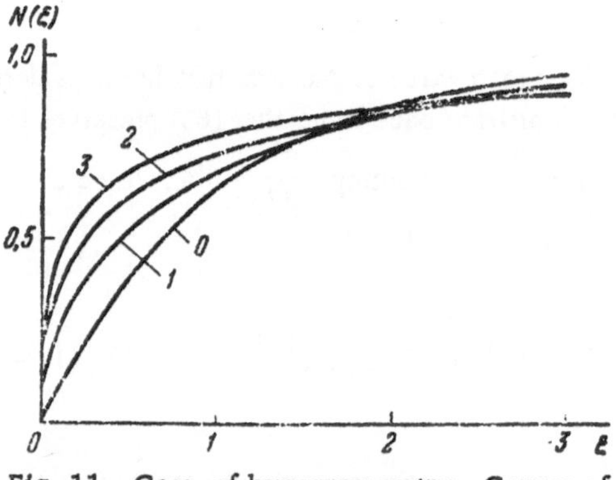

Fig. 11. Case of hypersymmetry. Curves of N(ζ) for different values of n.

Let us isolate an asymmetric group of atoms in the unit cell of a centrosymmetric crystal. Let us double it with the aid of a crystallographic inversion center — two groups of atoms will arise. In addition, let there be in the cell n − 1 noncrystallographic inversion centers. We shall create four groups of atoms with the first of them, with the second eight groups of atoms, etc. It is clear that, if the general number of inversion centers (including one crystallographic one) is n, then the number of asymmetric groups of atoms in such a cell will be 2^n. We shall call the structure of such a crystal n-centrosymmetric. For a crystal without hypersymmetry n = 1.

Let us isolate an asymmetric group of atoms in the unit cell of a crystal without an inversion center. Let us double it by a noncrystallographic translation — two groups of atoms will arise. Let us double these by a second translation vector — four groups of atoms will arise; the n-translation-vectors will create 2^n asymmetric groups. Let us call the structure of such a crystal n-parallel. A crystal without hypersymmetry has n = 0.

The following important theorem has been proved by the authors cited above. The intensity distribution of n-parallel and n-centrosymmetric crystals is identical for equal n. Thus, for example, the intensity distribution for a crystal of group P1 with two parallel molecules is identical in general with that for a crystal of group $P\bar{1}$ with one molecule in the general position; in a crystal of group $P2_1$, with two centrosymmetric molecules in general positions, and in one of group $P2_1/a$ with one molecule, the inversion center of which coincides with the crystallographic one; etc.

The distribution of structure amplitudes for hypersymmetry with the number n is derived from the formula

$$P(\hat{F})\,d\hat{F} = \frac{2\hat{F}d\hat{F}}{\pi^n \sum n_j^2} \int_0^{\frac{\pi}{2}} \cdots \int_0^{\frac{\pi}{2}} \left(\exp \frac{-\hat{F}^2 \sec^2\psi_1 \ldots \sec^2\psi_n}{2^n \sum n_j^2} \right) \times$$

$$\times \sec^2\psi_1 \ldots \sec^2\psi_n d\psi_1 \ldots d\psi_n, \qquad (63)$$

where ψ_n are the variables of integrations.

From (63), all the criteria of hypersymmetry can be computed — the same criteria as those analyzed in the preceding paragraph.

The values of $N(\zeta)$, i.e., the fraction of the magnitudes of $\hat{F}^2/\overline{\hat{F}^2}$ smaller than ζ, is given by the formula originating from (63):

$$N(\zeta) = \left(\frac{2}{\pi}\right)^{n-1} \int_0^{\frac{\pi}{2}} \cdots \int_0^{\frac{\pi}{2}} \operatorname{erf}\left[\left(\frac{\zeta}{2^n}\right)^{\frac{1}{2}} \sec\psi_2 \ldots \sec\psi_n\right] d\psi_2 \ldots d\psi_n. \quad (64)$$

From this follow the formulas analyzed previously; for a crystal with a complete loss of the inversion center

$$N_0(\zeta)=1-e^{-\zeta},$$

for a centrosymmetric crystal

$$N_1(\zeta)=\operatorname{erf}\left(\frac{\zeta}{2}\right)^{\frac{1}{2}},$$

and for a crystal with hypersymmetry with one supplementary inversion center

$$N_2(\zeta)=2\pi^{-1}\int_0^{\frac{\pi}{2}}\operatorname{erf}\left[\frac{1}{2}\zeta^{\frac{1}{2}}\sec\psi\right]d\psi.$$

One can verify that the last formula is identical with (62).

Figure 11 shows curves that make possible a comparison of these three simplest cases with that for n = 3.

Values of the numerical criteria we used in the preceding paragraph are also of interest. The ratio $\rho = |\overline{F}|^2/\overline{F^2}$ is expressed for hypersymmetry of the order n by the formula

$$\rho_n=2^{3n-2}\pi^{1-2n},$$

and the magnitude of the specific mean square deviation

$$v_n=2\left(\frac{3}{2}\right)^n-1.$$

A table of values of ρ_n and v_n appears as

n	0	1	2	3	4	5
ρ	$\frac{\pi}{2}=0.785$	$\frac{2}{\pi}=0.637$	0.516	0.418	0.339	0.275
v	1	2	$3\frac{1}{2}$	$5\frac{3}{4}$	9.12	14.2

It becomes obvious from this table that it is possible to prove hypersymmetry with certainty by computing the experimental magnitudes ρ and v.

Probably, however, such an analysis on the basis of the x-ray data may only ascertain the presence of some kind of hypersymmetry, since, besides the cases described, it is possible to suggest many other schemes of supersymmetry. However, if one possesses some crystallochemical data concerning the molecular framework, computations of ρ and $\underline{v}$, or of the structure of $N(\zeta)$, may serve as an important criterion for choosing one model from among several possible structures.

8. The Mean Value of the Structure Factor and the Problem of Finding the Unitary Amplitudes

In the preceding paragraphs we widely applied (39) which determines the mean value of the structure factor. There remain two questions to be answered: what is the accuracy of (39), and how may the individual structure amplitudes be found experimentally [5]

Let us begin with the question of the accuracy of (39). In a way, an answer has already been given above in Section 3, since the accuracy of (39) is determined by the degree of accuracy of fulfilling the Gaussian distribution, in other words, by the accuracy with which $\overline{\hat{F}^2}$ is equal to the second order moment of the distribution of structure amplitudes.

Nevertheless, in view of the importance of the question, we shall analyze it more directly.

Let us write $\hat{F}^2$ in the form

$$\hat{F}^2 = \left|\sum n_j e^{2\pi i \mathbf{H}\mathbf{r}_j}\right|^2 = \sum_j n_j^2 + \sum_{j\neq m}\sum n_j n_m e^{2\pi i \mathbf{H}\mathbf{r}_{jm}}, \tag{65}$$

where $\mathbf{r}_{jm} = \mathbf{r}_j - \mathbf{r}_m$.

[5] Let us note that (39) may be used in some cases to determine the coordinates of a heavy atom. It is not hard to see that the mean values of $\hat{F}^2_{hkl}$ over $\underline{h}$ and l for a fixed $\underline{k}$ will give, for example, for the group $P2_1/c$:

$$\overline{F^2_{k+l=2n}} = 8n_H^2\cos^2 ky_H + Nn_L^2;$$
$$\overline{F^2_{k+l\neq 2n}} = 8n_H^2\sin^2 ky_H + Nn_L^2.$$

Here n_H and n_L are the "weights" of the heavy and light atoms, N is the number of light atoms and y_H are the coordinates of the heavy atom. As we see, the experiment gives at once the magnitude of $\cos 2ky_H$.

It is known that the nodes of the reciprocal lattice, where the intensities can be established experimentally, lie within a sphere of some radius H_m. Let there be M nodes within this sphere. Then the mean value of the unitary structure factor derived from these M nodes will be equal[6] to

$$\langle \hat{F}^2 \rangle = \sum n_j^2 + \sum_j \sum_m n_j n_m \langle e^{2\pi i \mathbf{H} \mathbf{r}_{jm}} \rangle, \qquad (66)$$

where

$$\langle e^{2\pi i \mathbf{H} \mathbf{r}_{jm}} \rangle = \frac{1}{M} \sum e^{2\pi i \mathbf{H} \mathbf{r}_{jm}} \approx \frac{1}{\tau_M} \int e^{2\pi i \mathbf{H} \mathbf{r}_{jm}} \, d\tau.$$

The substitution of the sum by the integral produces a small error, which is smaller the larger M is (usually M > 1000). The integral must be taken over a sphere of radius H_0; τ_M is the volume of this sphere. The computation of the integral

$$\frac{1}{\tau_M} \int e^{2\pi i \mathbf{H} \mathbf{r}_{jm}} d\tau = \frac{1}{\tau_M} \int_0^{2\pi} \int_0^{\pi} \int_0^{H_0} e^{2\pi i H r_{jm} \cos\alpha} H^2 \sin\alpha \, d\alpha \, d\varphi \, dH$$

gives

$$\langle e^{2\pi i \mathbf{H} \mathbf{r}_{jm}} \rangle = \frac{6\pi}{(2\pi H_0 r_{jm})^3} [\sin 2\pi H_0 r_{jm} - 2\pi H_0 r_{jm} \cos 2\pi H_0 r_{jm}].$$

The value of H_0 in the usual experiment is somewhat larger than 1 A^{-1}. The magnitudes r_{jm} are larger than 1.5 A and reach tens of angstroms. The values of the trigonometric functions are quite random for different r_{jm}, since in a usual structure, the arguments change by tens of complete circumferences corresponding to the variation in r_{jm}. We can only evaluate the maximum value of the average under consideration. In any case, the absolute magnitude of

$$\langle e^{2\pi i \mathbf{H} \mathbf{r}_{jm}} \rangle < \frac{6\pi}{(2\pi H_0 r_{jm})^3} + \frac{6\pi}{(2\pi H_0 r_{jm})^2}.$$

Even for $r_{jm} = 1.5$ and, consequently, $2\pi H_0 r_{jm} \approx 10$ the expression

$$\langle e^{2\pi i \mathbf{H} \mathbf{r}_{jm}} \rangle < 0.1.$$

[6] < > is the symbol of the average, as is the bar above.

But for the majority of the vectors $\mathbf{r}_{jm}$ the expression under consideration will be much smaller than 0.1. In addition, taking into account that the signs of $\langle e^{2\pi i \mathbf{H}\mathbf{r}_{jm}} \rangle$ are quite random, and that the number of distances even for a simple structure is measured in hundreds, it becomes clear that the double sum in (65) is negligibly small in comparison with $\sum n_j^2$ and, consequently (39) is fulfilled.

From the examples given [29], it is evident that the last equation is fulfilled with great accuracy

	$\hat{F}^2_{calc}$	$\hat{F}^2_{meas}$
Dicyanamide, for the zone h0l	0.289	0.309
Melamine for the zone h0l	0.235	0.246
Melamine for the zone 0kl	0.167	0.167
Cyanuric Triazide for the zone hk0	0.182	0.183

The agreement is quite remarkable, if one takes into account that in computing the averages an insignificant number of reflections was used.

For what comes later it is also necessary to analyze the case when the mean value $\langle \hat{F}^2 \rangle$ is computed not for the whole interference field, but for a narrow spherical layer with the radii $H_0 \pm \Delta H$. For such a layer the average value of $\langle \hat{F}^2 \rangle$ may be equated to the mean value on a surface of radius H_0:

$$\langle e^{2\pi i \mathbf{H}\mathbf{r}_{jm}} \rangle_{\text{layer}} = \frac{1}{4\pi H_0^2} \int e^{2\pi i \mathbf{H}\mathbf{r}_{jm}}\, d\sigma.$$

Computation gives

$$\langle e^{2\pi i \mathbf{H}\mathbf{r}_{jm}} \rangle_{\text{layer}} = \frac{\sin 2\pi H_0 r_{jm}}{2\pi H_0 r_{jm}}.$$

In this case (66) acquires the form

$$\langle \hat{F}^2 \rangle = \sum_j n_j^2 + \sum_{i \neq j}\sum n_i n_j \cdot \frac{\sin 2\pi H_0 r_{ij}}{2\pi H_0 r_{ij}}. \qquad (67)$$

It is not hard to see that the double sum cannot give any essential contribution to the mean value of the unitary structure factor.

The shape of the function $\frac{\sin x}{x}$ is well known: From a value of unity the function decreases to zero when $x = \pi$, and then it oscillates about zero, giving weak decreasing maxima.

If all the arguments entering the double sum $2\pi H r_{ij} > 2$, i.e., all the interatomic distances

$$r_{ij} > \frac{1}{3} d_{hkl}, \qquad (68)$$

then the double sum will be very close to zero, because of the oscillating character of $\frac{\sin x}{x}$.

The smallest distances r_{ij} between the valence-bound atoms are magnitudes of the order of 1.5 A. If there are N atoms in the cell, then there will be $N(N-1)$ magnitudes r_{ij} in all, of which about 2N or 3N will be among the shortest of them. It is clear that these interatomic distances will play a small part. For $d_{hkl} < 3$ A the double sum over the atoms of the cell will have quite a negligible magnitude and (39) preserves its form for a spherical layer as well.

Thus, (39) is applicable to any spherical layer which includes a statistically sufficient number of nodes. Only the regions of reciprocal space adjoining the origin of coordinates will, as was known beforehand, comply poorly with (39); first, because in this region there will be nodes with maximum d_{hkl} second, because the radius of the sphere will be large, but the number of nodes in the sphere small. The best results from (39) will occur for large H_0.

Equation (39) holds for an overwhelming majority of crystals. Crystals of the globular protein type are an exception, since they have reflections only for planes with $d > 2$ A. However, as Harker [30] has pointed out, here, too statistical considerations have a certain advantage. It is not hard to see that in any case

$$\langle F^2 \rangle = \sum \overline{|F_j(\mathbf{H})|^2},$$

where $F_j(\mathbf{H})$ is the structure amplitude of a globule. Some atomic grouping must be chosen as the globule [large enought in the sense of inequality (68)], which may be considered as the brick from which the protein crystal is constructed.

Since

$$F = \hat{f}\hat{F}\sum Z_j,$$

where $\hat{f}$ is a unitary atomic factor, then, for a spherical layer with a mean radius H, one can write

$$\overline{F^2} = \hat{f}^2_{H_0}\left(\sum Z_j\right)^2 \cdot \overline{\hat{F}^2},$$

or, since

$$\hat{F}^2 = \sum n_j^2 = \frac{\sum Z_j^2}{\left(\sum Z_j\right)^2}, \tag{69}$$

$$\overline{F^2} = \hat{f}^2_{H_0}\sum Z_j^2$$

or, according to the definition of the unitary structure amplitude

$$\overline{F^2_{H_0}} = \sum f_j^2(H_0). \tag{70}$$

This important formula was first derived by Wilson.

Let us now consider the method of computing unitary structure amplitudes from the experimental data.

Equation (39) again lies at its basis.

Using the values of the mean structure factor (not unitary) for all shells, we can construct a smooth $\langle F^2 \rangle$ curve as a function of H on some arbitrary scale, which is, however, the same one in which the values of F^2 are measured.

If we succeed in extrapolating this curve with certainty to H = 0:

$$\langle F^2 \rangle_{H=0} = \sum Z_j^2, \tag{71}$$

then a conversion factor is obtained to absolute (electron) units for the measured structure amplitudes, and the $\hat{f}^2$-curve can be constructed on the basis of (70).

Exact extrapolation, however, is impossible in the majority of cases. Therefore (70) usually determines an approximate value of the conversion factor and establishes the $\hat{f}^2$-curve except for a constant factor which may differ from unity even by 0.2-0.3.

However, (39) has the advantage that with its aid the transition from F_{meas} to $\hat{F}_{meas}$ is possible without the knowledge, not only of the f-curves, but also without information concerning the $\hat{f}$-curve, i.e., without using (71).

In fact, dividing (39) by (70) and extracting the root, we obtain (see [12])

$$\hat{F}=\frac{F}{\sqrt{<F^2>}}\frac{\sqrt{\sum Z_j^2}}{\sum Z_j}. \qquad (72)$$

In the case of identical atoms (72) becomes

$$\hat{F}=\frac{F}{\sqrt{<F^2>}}\frac{1}{\sqrt{N}}. \qquad (73)$$

Equation (72) shows the value of the unitary structure amplitudes – these quantities may be obtained experimentally solely on the basis of knowing the atomic composition of the crystal. Structure analysis without any information whatsoever concerning the f-curves[6] becomes possible.

To establish a structure from the experimental data, the unitary structure amplitudes are more convenient to use than the quantities F.

Agreement between the $\hat{F}_{meas}$ [from (72)] and the $\hat{F}_{calc} = 2\sum_{1}^{N/2}{}' n_j \cos\alpha$ is the criterion of the accuracy of the structure.

Let us note that the unitary $\hat{f}$-curve is determined by (69) and (70) and not by the formulas of the mean $\hat{f}=\frac{\sum f_j}{\sum Z_j}$ or of the weighted average $\hat{f}=\frac{\sum f_j Z_j}{\sum Z_j^2}$. By determining the $\hat{f}$-curve from formula

$$\hat{f}^2=\sum f_j^2/\sum Z_j^2 \qquad (74)$$

[6]By giving the results of measurement on an absolute scale, the relationship $\frac{1}{V}\sum_{hkl} F_{hkl}^2=\int_V \rho^2 dv$ may be useful, where V is the volume of the cell. This equation follows directly from (1.15) see G. Kartha [31] and B. K. Vainshtein [32].

it is assumed that the best structure is the one in which the mean square values of the measured and the computed structure amplitudes coincide.

To terminate the discussion concerned with the use of unitary structure amplitudes, one must answer the question as to what errors may be introduced into the structure analysis by the approximate assumption which postulates the similarity of the f-curves. Let $\hat{f}$ be derived from (74). The value of the jth atomic factor can be written in the form

$$f_j = (\hat{f} + \Delta\hat{f}_j)\, Z_j.$$

Here $\Delta\hat{f}_j$ is the deviation of the correct atomic factor from the unitary one. Then

$$F = \sum f_j e^{2\pi i \mathbf{H}\mathbf{r}_j} = \sum Z_j (\hat{f} + \Delta\hat{f}_j)\, e^{2\pi i \mathbf{H}\mathbf{r}_j} =$$
$$= \sum Z_j \hat{f} e^{2\pi i \mathbf{H}\mathbf{r}_j} + \sum Z_j\, \Delta\hat{f}_j e^{2\pi i \mathbf{H}\mathbf{r}_j}$$

or, dividing by $\hat{f} \sum Z_j$, we obtain

$$\frac{F}{\hat{f} \sum Z_j} = \sum n_j e^{2\pi i \mathbf{H}\mathbf{r}_j} + \sum \frac{\Delta\hat{f}_j}{\hat{f}}\, n_j e^{2\pi i \mathbf{H}\mathbf{r}_j}\,. \qquad (75)$$

Neglecting $\frac{\Delta\hat{f}_j}{\hat{f}}$ we arrive at (72). If the second sum becomes comparable in magnitude with the first one, then the values of $\hat{F}$ computed experimentally according to (72) will diverge from the values of $\hat{F}$ computed for the structure by the formula $\hat{F} = 2\sum n_j \cos\alpha_j$. Moreover, if the second sum is larger than the first, then the unitary structure amplitude $\hat{F}$ may have a sign opposite to that of the structure amplitude F (antagonistic reflections) [see [4], p. 473 and ch. VI).

From the definition of the $\hat{f}$-curve it is clear that $\Delta\hat{f}_j$ can have both positive and negative values. From the same definition, it follows that the $\hat{f}$-curve will be closest to the $\frac{f_j}{z_j}$-curve of the heavy atom. Therefore, speaking generally, the values of the second sum in (75) cannot be large.

However, accidental lapses are quite possible (let us imagine, for example, that the atoms with $\Delta\hat{f}_j$ of the same sign have exponential factors close to zero).

To get an idea of the limits of applicability of using unitary structure amplitudes, let us examine the mean square (75), assuming the values of $\mathbf{Hr}_j$ to be uniformly distributed over the trigonometric circle:

$$\overline{\hat{F}^2_{\text{meas}}} = \left\langle \left(\frac{F}{\hat{f}\sum Z_j} \right)^2 \right\rangle = \left\langle \left(\sum \left(1 + \frac{\Delta\hat{f}_j}{\hat{f}}\right) n_j e^{2\pi i \mathbf{Hr}_j} \right)^2 \right\rangle = \\ = \sum \left(1 + \frac{\Delta\hat{f}_j}{\hat{f}}\right)^2 n_j^2$$

and, assuming positive and negative values of $\Delta\hat{f}$ to be equally probable:

$$\overline{\hat{F}^2_{\text{meas}}} = \left\langle \left(\frac{F}{\hat{f}\sum Z_j} \right)^2 \right\rangle = \sum \left[1 + \left(\frac{\Delta\hat{f}_j}{\hat{f}} \right)^2 \right] n_j^2 .$$

If the deviations from the $\hat{f}$-curve are small, then

$$\overline{\hat{F}^2_{\text{meas}}} = \overline{\hat{F}^2_{\text{calc}}} = \sum n_j^2 .$$

We may consider that the applicability of the assumption of the similarity of the f-curves is limited to the requirement

$$\sum \left(\frac{\Delta\hat{f}_j}{\hat{f}} \right)^2 n_j^2 < \sum n_j^2 .$$

For atoms with neighboring atomic numbers this condition acquires the form

$$\left(\frac{\Delta\hat{f}}{\hat{f}} \right)_{\text{av}} < 1. \qquad (76)$$

This condition is easily fulfilled for small and medium scattering angles. But for large values of H (76) may not be fulfilled, as is shown by experimental examples.

Condition (76) may be considered very mild, and as justifying fully the use of unitary structure amplitudes in structure analysis. Considerable deviations of $\hat{F}_{\text{meas}}$ from $\hat{F}_{\text{calc}}$, due to the incorrectness of the assumption that the f-curves are similar, will occur only for that fraction of the reflections having large values of the scattering angle. Deviations which lead to a divergence between the signs of the $\hat{F}$ and the F magnitudes will occur even more rarely – only for those reflections for which $\hat{F} \ll 1$.

At the same time, only this last circumstance has meaning. As we shall see below, the determination of the signs of the F's is the goal of the transition to unitary structure amplitudes, and not the comparison between $\hat{F}_{meas}$ and $\hat{F}_{calc}$. Knowing F by the order of magnitude (the strongest, strong, etc.) is quite sufficient for analysis.

For us only the following question is of importance: do the existing objective methods (i.e., not derived from the experimental structure) for determining the signs of $\hat{F}$ include the assumption of the similarity of the f-curves? If they do, then (76) puts a limit to the accuracy of structure determination; if they do not, then the divergence of $\hat{F}_{meas}$ from $\hat{F}_{calc}$ has no meaning in general and is of no interest.

We shall return to this question in chapter VI.

9. Concerning the Possibility of Determining the Space Group from Intensity Statistics

If one is guided by the Laue symmetry of the x-ray patterns, and by the presence or absence of systematic extinctions of reflections, then, as is known, it becomes possible to classify crystals into 122 x-ray groups.

The statistical rules discussed above allow one in principle to "split" an x-ray group into the space groups of which it is composed. To do this, it is necessary to calculate the statistical characteristics of the general reflections and the reflections of zones and rows. For each group of reflections $\bar{\hat{F}}^2$, $N(\zeta)$, ρ and $\underline{v}$ may be determined experimentally.

We shall show that a knowledge of $\bar{\hat{F}}^2$ and of the function $N(\zeta)$ (or ρ, or $\underline{v}$) allow one in principle (we emphasize this word) to solve the problem of determining the space group.

Let us consider, for example the x-ray group with the diffraction symbol $2/mP-/c$. It includes two space groups: $C_{2h}^4 = P2/c$ and $C_s^2 = Pc$. The values of $\bar{\hat{F}}^2$ for reflections of all types are identical. Statistical differences between these two space groups are that for general reflections, the first space group will give $\rho = 0.637$ and $v = 2$, and the second space group will give $\rho =$ $= 0.785$ and $v = 1$.

The determination of the inversion center by the statistical method in this and similar cases has a disadvantage shared by all physical methods of studying this symmetry element. One may prove the absence of an inversion center, but not its presence. In fact, the result $\rho = 0.785$ and $v = 1$ has a unique significance — the center of inversion is absent. The result $\rho = 0.637$ and $v = 2$ may occur, as we have observed, for a parallel arrangement of two groups of atoms in the cell without a center of inversion. Furthermore, a partial loss of a center of inversion is always possible, which may bring one to figures close to $\rho = 0.637$ and $v = 2$; however, the crystal will not have a true center of inversion.

Nevertheless, the statistical treatment, especially using reflections of the general type (of which there are ordinarily the order of a thousand) is unquestionably useful, and should be carried out in all cases when the presence or absence of an inversion center is debatable and does not follow from packing considerations.

Let us examine some more examples of space group determinations by the statistical method. The diffraction group with the symbol mmmPna includes two space groups: D_{2h}^{16} = Pnam and C_{2v}^{9} = Pna2. Since the distinction is in the inversion center, one should attempt to establish the space group by calculating ρ and $\underline{v}$, just as in the preceding example. Let us note that in this case, the groups are also distinguished by the values of $\overline{\hat{F}^2}$ for various types of indices, namely for the group D_{2h}^{16} in the orientation given above.

$$\overline{\hat{F}^2_{hk0}} = \overline{\hat{F}^2_{0kl}} = \overline{\hat{F}_{h0l}} = \frac{1}{2}\overline{\hat{F}^2_{hkl}};$$

while for the group C_{2v}^{9}

$$\overline{\hat{F}^2_{hk0}} = \frac{1}{2}\overline{\hat{F}^2_{0kl}} = \frac{1}{2}\overline{\hat{F}^2_{h0l}} = \frac{1}{2}\overline{\hat{F}^2_{hkl}}.$$

In addition, let us examine the x-ray group mmmPc— —. It comprises the three space groups D_{2h}^{5} = Pcmm; C_{2v}^{2} = Pcm2$_1$ and C_{2v}^{4} = Pc2m. Investigation of the inversion center using the general indices may distinguish the first from one of the two groups of class C_{2v}. One may look for the distinction between these two groups by analyzing the centrosymmetricity of the projections. In C_{2v}^{2} the projection ab has a center of symmetry, in C_{2v}^{4} – the projection ac. Consequently, the values of ρ and $\underline{v}$ calculated for these zones will be $\rho = 0.785$ in C_{2v}^{2}, and $v = 1$ for the indices h0l and 0kl, in C_{2v}^{4} for the indices hk0 and 0kl. The values $\rho = 0.637$ and $\underline{v} = 2$ will arise correspondingly for hk0 in C_{2v} and for h0l in C_{2v}^{4}. Differences in $\overline{\hat{F}^2}$ in principle allow one to distinguish between all the three groups. In D_{2h}^{5}

$$\overline{\hat{F}^2_{hk0}} = \overline{\hat{F}^2_{h0l}} = \overline{\hat{F}^2_{0kl}} = \frac{1}{2}\,\overline{\hat{F}^2_{hkl}};$$

in C_{2v}^2

$$\overline{\hat{F}^2_{hk0}} = \frac{1}{2}\,\overline{\hat{F}^2_{h0l}} = \frac{1}{2}\,\overline{\hat{F}^2_{0kl}} = \frac{1}{2}\,\overline{\hat{F}^2_{hkl}}$$

and in C_{2v}^4

$$\overline{\hat{F}^2_{h0l}} = \frac{1}{2}\,\overline{\hat{F}^2_{hk0}} = \frac{1}{2}\,\overline{\hat{F}^2_{0kl}} = \frac{1}{2}\,\overline{\hat{F}^2_{hkl}}.$$

Examining in detail all the x-ray groups we come to the conclusion that the x-ray experiment fails to distinguish only the enantimorphous pairs of space groups, and also the groups I222 from $I2_12_12_1$ and I23 and $I2_13$.

Complete tables for space group determinations by the method just indicated have been drawn up by Rogers [33].

The statistical method enriches considerably the possibilities of establishing the symmetry of a crystal by x-ray methods. However, it should be remembered that various kinds of hypersymmetry and approximation in realizing any element of symmetry can always lead one into error. Here the assertion that we have made with respect to the center of symmetry is correct: The absence of this element of symmetry can be proved rigorously, but its presence is proved only by a favorable termination of the complete structure analysis.

10. The Distribution Function of the Third-Order Structure Product

Let us examine ([21] and [20] page 225) in this paragraph a centrosymmetric crystal for which the unitary structure amplitude has the form

$$\hat{F}_H = 2\sum_{1}^{N/2} n_j \cos\alpha_j,$$

and the structure product of the third order is equal to

$$W = \hat{F}_H \hat{F}_K \hat{F}_{H+K} = 8 \sum_{j,\,r,\,p=1}^{N/2} n_j n_r n_p \cos\alpha_j \cos\beta_r \cos(\alpha_p + \beta_p). \quad (77)$$

The arguments α and β are considered as random quantities, and the distribution function of W is sought, considering all the addends of the triple sum to be independent.

The average value of

$$q_{jrp} = \cos \alpha_j \cos \beta_r \cos (\alpha_p + \beta_p),$$

obviously, is equal to

$$\bar{q}_{jrp} = \frac{1}{4} \tag{78}$$

when $j = r = p$ and $q_{jrp} = 0$; in the remaining cases, consequently,

$$\overline{W} = \sum_{j=1}^{N} n_j^3 \tag{79}$$

or for identical atoms in the cell,

$$\overline{W} = \frac{1}{N^2}.$$

Thus, for the structure product W under study, the distribution function (35) is correct, whereupon

$$U = W - \overline{W} = W - \sum_{j=1}^{N} n_j^3$$

or, for identical atoms,

$$U = W - \frac{1}{N^2}.$$

We write the expression for U in the form (10):

$$U = \sum_{jrp} u_{jrp}, \tag{80}$$

where

$$u_{jrp} = 8n_j n_r n_p [\cos \alpha_j \cos \beta_r \cos (\alpha_p + \beta_p) - \bar{q}_{jrp}]. \tag{81}$$

The mth moment of the quantity u_{jrp}, according to (19), is the mean of the mth power of u_{jrp}, i.e.,

$$a_{(jrp)m} = 8^m n_j^m n_r^m n_p^m \overline{[\cos \alpha_j \cos \beta_r \cos (\alpha_p + \beta_p) - \bar{q}_{jrp}]^m}$$

or for atoms of one kind

$$a_{(jrp)m} = 8^m \frac{1}{N^{3m}} \overline{[\cos\alpha_j \cos\beta_r \cos(\alpha_p + \beta_p) - \bar{q}_{jrp}]^m}.$$

The average magnitude of the square brackets has only 2 values: one for j = r = p, the other for cases when the indices are not equal. Let us introduce the notations

$$\varepsilon_m^{123} = \overline{[\cos\alpha_j \cos\beta_r \cos(\alpha_p + \beta_p)]^m};$$

$$\varepsilon_m^{111} = \overline{\left[\cos\alpha_j \cos\beta_j \cos(\alpha_j + \beta_j) - \frac{1}{4}\right]^m}.$$

For odd m $\epsilon_m^{123} = 0$, for even

$$\varepsilon_m^{123} = \overline{[\cos^m \alpha]} = \left[\frac{(m-1)!!}{m!!}\right]^3 \tag{82}$$

For ϵ_m^{111} we have

$$\varepsilon_m^{111} = \sum_k \binom{m}{k} \overline{[\cos\alpha \cos\beta \cos(\alpha+\beta)]^k} \left(-\frac{1}{4}\right)^{m-k}. \tag{83}$$

In addition, since $\overline{\cos^p \alpha \sin^q \alpha} = \frac{(p-1)!!(q-1)!!}{(p+q)!!}$, we obtain for the average under the summation sign with an accuracy of the order of 1%:

$$\overline{[\cos\alpha \cos\beta \cos(\alpha+\beta)]^k} \approx \left[\frac{(2k-1)!}{2k!}\right]^2 1.15.$$

According to (82) and (83), for different j, r, p only two different values of the mth moment will occur: a_m^{111} – for j = r = p and a_m^{123} for the remaining cases:

$$a_m^{111} = \left(\frac{2}{N}\right)^{3m} \varepsilon_m^{111}; \quad a_m^{123} = \left(\frac{2}{N}\right)^{3m} \varepsilon_m^{123}, \tag{84}$$

The number of addends in (80) with the indices j = r = p for the case under study is equal to N/2, the number of the remaining addends is $\left(\frac{N}{2}\right)^3 - \frac{N}{2}$. Consequently, the formulas for the moments of U computed according to (25) and (26), have the form

$$\left.\begin{aligned} A &= \frac{N}{2}\,a_2^{111} + \left(\frac{N^3}{8} - \frac{N}{2}\right) a_2^{123}; \\ A_m &= \frac{N}{2}\,\alpha_m^{111} + \left(\frac{N^3}{8} - \frac{N}{2}\right) \alpha_m^{123}, \end{aligned}\right\} \tag{85}$$

where the α_m are computed from the corresponding a_m according to (23).

For the moments of the second order we have

$$a_2^{111} = \frac{3}{32}, \quad a_2^{123} = \frac{1}{8}, \quad \alpha_2^{111} = \left(\frac{2}{N}\right)^6 \cdot \frac{3}{32}, \quad \alpha_2^{123} = \left(\frac{2}{N}\right)^6 \cdot \frac{1}{8}.$$

Hence, the basic parameter of the distribution function — the mean square of (80) is

$$A = \left(\frac{2}{N}\right)^5 \cdot \frac{3}{32} + \left(\frac{2}{N}\right)^3 \cdot \frac{1}{8} - \left(\frac{2}{N}\right)^5 \cdot \frac{1}{8},$$

i.e.,

$$A = \frac{1}{N^3} - \frac{1}{N^5} \approx \frac{1}{N^3}.$$

The computation of the first few coefficients A_m is set forth in Table 3. Consequently, the coefficients β of (29) acquire the form

$$\beta_3 = -0.0039\left(\frac{2}{N}\right)^8, \quad \beta_4 = 2.5 \cdot 10^{-4}\left(\frac{2}{N}\right)^9 - 4.4 \cdot 10^{-4}\left(\frac{2}{N}\right)^{11},$$

$$\beta_5 = -9 \cdot 10^{-5}\left(\frac{2}{N}\right)^{14}, \quad \beta_6 = 1.4 \cdot 10^{-5}\left(\frac{2}{N}\right)^{15} - \frac{1}{2}\,(0.0039)^2\left(\frac{2}{N}\right)^{16} -$$

$$-1.0 \cdot 10^{-5}\left(\frac{2}{N}\right)^{17} \text{ etc.}$$

The distribution function (35) of the structure product, which we shall now denote by X, is equal to

$$f(X)=\frac{1}{\sqrt{2\pi/N^3}}e^{-\frac{\left(X-\frac{1}{N^2}\right)^2}{2/N^3}}\Big\{1-\left[0.0039\cdot\left(\frac{2}{N}\right)^8\right]\left[u-\frac{1}{3}u^3\right]3N^6+$$
$$+\left[2.5\cdot10^{-4}\left(\frac{2}{N}\right)^9-4.4\cdot10^{-4}\left(\frac{2}{N}\right)^{11}\right]\left[1-2u^2+\frac{1}{3}u^4\right]3N^6-$$
$$-\left[9\cdot10^{-5}\left(\frac{2}{N}\right)^{14}\right]\left[u-\frac{2}{3}u^3+\frac{1}{15}u^5\right]15N^9+\left[1.4\cdot10^{-5}\left(\frac{2}{N}\right)^{15}-\right.$$
$$-0.8\cdot10^{-5}\left(\frac{2}{N}\right)^{16}-1\cdot10^{-5}\left(\frac{2}{N}\right)^{17}\Big]\Big[1-3u^2+u^4-$$
$$\left.-\frac{1}{15}u^6\right]15N^9\ldots\Big\}, \tag{86}$$

where $u=\left(X-\frac{1}{N^2}\right)/\sqrt{A}$.

TABLE 3. The First Few Coefficients A_m

m	ε_m^{111}	ε_m^{123}	A_m
3	0.023	0	$-\left(\frac{2}{N}\right)^8\cdot 0.0039$
4	0.022	$-\left(\frac{3}{8}\right)^3=0.053$	$\left(\frac{2}{N}\right)^9\cdot 2.5\cdot10^{-4}-4.4\cdot10^{-4}\left(\frac{2}{N}\right)^{11}$
5	0.011	0	$-\left(\frac{2}{N}\right)^{14}\cdot 9\cdot 10^{-5}$
6	0.008_6	$\left(\frac{15}{48}\right)^3=0.031$	$\left(\frac{2}{N}\right)^{15}\cdot 1.4\cdot10^{-5}-1.0\cdot10^{-5}\cdot\left(\frac{2}{N}\right)^{17}$
7	0.003_3	0	etc.
8	0.003_3	0.020	
9	0.002_8	0	
10	0.001_7	0.014	

The series in braces rapidly converges in all cases, with the exception of the case of one atom in a general position (N = 2).

Even for N = 8 the quantity in braces is in practice equal to unity for all values of u that may be encountered experimentally.

The example analyzed is quite sufficient for the following conclusion: when the correlation between the separate components of the structure product is neglected, its distribution function is approximated quite accurately by a Gaussian function.

The effect of correlation between the addends (which is very substantial) is analyzed in the following chapter.

11. The Gaussian Distribution of the Structure Products[7]

The conclusion of the preceding paragraph allows us to apply the Gaussian distribution

$$\varphi(X) = \frac{1}{\sqrt{2\pi D}} e^{-\frac{(x-\bar{x})^2}{2D}}. \tag{87}$$

to structure products of any order [the higher the order, the more rapidly the series (35) converges] and for any symmetry (concerning the effect of symmetry see below). Equation (87) shows the basic cause of our interest in structure products: the probability that X is positive is greater than the probability of its being negative, since $\bar{X}$ is always larger than zero. Let us emphasize once more that the Gaussian distribution is obtained on the assumption that correlation between the components of structure products is absent.

We shall see below that those quantities X which are, according to (87) improbable do not obey this distribution (i.e., $|X|$ are ten times greater than the mean square; see pages 177 and 202). However, in this case it turns out that the probability of a positive X approaches unity even more closely, following another law.

The magnitudes of the mean value of the structure product $\bar{X}$ and of the dispersion D may be computed for each concrete case. We shall prove some general theorems for these parameters. Let us write the expression for the structure amplitude in the form

$$\hat{F} = z \sum_{j}^{N/z} n_j \psi_j (H), \tag{88}$$

where z is the number of equivalent atoms, and let us compute the mth order parameters of the structure product distribution

[7][34]; see also page 121.

$$X = z^m \sum_{j_1, j_2, \ldots, j_m}^{N/z} n_{j_1} n_{j_2} \ldots n_{j_m} \psi_{j_1}(H_1) \psi_{j_2}(H_2) \ldots \psi_{j_m} \times$$
$$\times (H_1 + H_2 + \ldots + H_{m-1}). \quad (89)$$

Let us begin with the mean value. In averaging over the arguments of the trigonometric functions ψ_j all the addends become zero, with the exception of those for which all the indices of ψ_j are identical. Therefore

$$\overline{X} = g \sum_j^N n_j^m, \quad (90)$$

where

$$g = z^{m-1} \overline{\psi_j(H_1) \psi_j(H_2) \ldots \psi_j(H_1 + H_2 + \ldots + H_{m-1})}. \quad (91)$$

For the case of identical atoms

$$\overline{X} = g \frac{1}{N^{m-1}}. \quad (92)$$

We can consider the distribution function of a structure product with definite indices $H_1, H_2 \ldots H_{m-1}$, assuming the atomic coordinates to be independent, random magnitudes. Obviously the distribution parameters will depend on whether the indices belong to a zone, a row, or are of the general type, or on whether any of the indices are equal to one another, and, of course, on the degree of degeneracy of the structure product.

Let us limit ourselves to the lowest symmetries, i.e., let us consider the cases where $\psi(H) = \cos\alpha$, $\psi(H) = \cos\alpha \cos\beta$ and $\psi(H) = \cos\alpha \cos\beta \cos\gamma$.

Let us use Euler's formulas for cosines:

$$\cos\alpha_1 \cos\alpha_2 \ldots \cos(\alpha_1 + \alpha_2 + \ldots + \alpha_{m-1}) =$$
$$= \frac{1}{2^m}(e^{i\alpha_1} + e^{-i\alpha_1})(e^{i\alpha_2} + e^{-i\alpha_2}) \ldots (e^{i(\alpha_1+\alpha_2+\ldots+\alpha_{m-1})} + e^{-i(\alpha_1+\alpha_2+\ldots+\alpha_{m-1})}) \quad (93)$$

From this way of writing, it is quite apparent that the product of the cosines that interests us may be represented in the form

$$\frac{1}{2^m}\left(2 + 2\sum\right), \quad (94)$$

where $\sum$ is a sum of cosines of different linear combinations of the arguments α.

From (93) it is obvious that the equality of different α's to one another will not affect the form of (94); only the arguments of the individual cosines that enter into $\sum$ will change. Thus, a dengeneracy of the structure product is not important for X.

Now it is not difficult to calculate (91). For triclinic symmetry

$$g = 2^{m-1}\frac{1}{2^m}\left(2+2\overline{\sum}\right) = 1,$$

for monoclinic

$$g = 4^{m-1}\cdot\frac{1}{2^m}\left(2+2\overline{\sum}_1\right)\frac{1}{2^m}\left(2+2\overline{\sum}_2\right) = 1.$$

We have the same for orthorhombic symmetry also.

Thus, it has been proved that in these cases g = 1. Therefore, for a structure product of the $\underline{m}$th order, irrespective of the degree of degeneracy,

$$\overline{X} = \sum_{j=1}^{N} n_j^m \quad \text{or} \quad \overline{X} = \frac{1}{N^{m-1}}, \tag{95}$$

if all the indices are of the general type.

How does a choice of indices with zeros affect X? From the sense of the computation it is clear that only one thing is essential: by what is the structure amplitude represented – by a cosine or by a product of two or three cosines. If, let us say, only the indices $h0l$ of the zone that passes through the axis of symmetry of a monoclinic crystal enter the structure product, then

$$g = 4^{m-1}\frac{1}{2^m}\left(2+2\overline{\sum}_1\right), \text{ i.e., } g = 2^{m-1}.$$

A more general conclusion is the following. It is essential to convert those indices into zero which reduce $\cos\alpha$ to unity.

If there are p such reductions to zero, then the denominator in (93) becomes equal to 2^{m-p}.

From this it follows that if in a structure product there are, in general, κ intrinsic zeros among the indices hk$l$ then

$$g = 2^{\kappa}. \tag{96}$$

We see that there is no explicit dependence of the mean value of the structure product on symmetry.

Let us pass over to the computation of the dispersion. Since $D = \overline{X^2} - \overline{X}^2$, we have to determine the mean square value of the structure product. Squaring (89) we obtain

$$X^2 = z^{2m} \sum n_{j_1} n_{j_1'} n_{j_2} n_{j_2'} \ldots \psi_{j_1}(H_1)\,\psi_{j_1}'(H_1)\,\psi_{j_2}(H_2)\,\psi_{j_2}'(H_2)\ldots$$
$$\ldots \psi_{j_m}(H_1 + H_2 + \ldots + H_{m-1})\,\psi_{j_m}'(H_1 + H_2 + \ldots + H_{m-1}). \tag{97}$$

In averaging (97) only those terms in the sum will remain for which either $j_1 = j_1'$, $j_2 = j_2'$, ..., $j_m = j_m'$, or $j_1 = j_2 = \ldots = j_m$, and simultaneously $j_1' = j_2' = \ldots = j_m'$.

Again using Euler's formulas (93), we obtain for the first type of addends:

$$\sum_{j_1, j_2 \ldots}^{N/z} n_{j_1}^2 n_{j_2}^2 \ldots \prod_{i=1}^{\varepsilon} \Big[\left(2 + e^{2i\alpha_{j_1}^i} + e^{-2i\alpha_{j_1}^i}\right)\left(2 + e^{2i\alpha_{j_2}^i} + e^{-2i\alpha_{j_2}^i}\right)\ldots$$
$$\ldots \left(2 + e^{2i\left(\alpha_{j_1}^i + \alpha_{j_2}^i + \ldots\right)} + e^{-2i\left(\alpha_{j_1}^i + \alpha_{j_2}^i \ldots\right)}\right)\Big],$$

where ϵ = 1, 2, 3 respectively for a triclinic, monoclinic, and orthorhombic crystal.

The factor $z^{2m} = 2^{2m\epsilon}$ was reduced by the same divisor $\left(\frac{1}{2^m}\right)^{2\epsilon}$ which is obtained by substituting in (93).

From the whole sum, the sum $\left(\sum_j^{N/z} n_j^2\right)^m 2^{m\epsilon}$ is isolated at once. Moreover, in averaging this does not become equal to zero, and also the additional

terms with $j_1 = j_2 = \ldots = j_m$, give the addend $2^z \sum_j^{N/z} n_j^{2m}$.

Also, as we have already indicated, those terms of the sum (97) do not disappear in averaging for which simultaneously $j_1 = j_2 \ldots = j_m$ and $j_1' = j_2' = \ldots = j_m'$. Each addend of this type has the form

$$z^2 n_j^m n_k^m g^2.$$

Thus, for the mean square value of the structure product we have finally

$$\overline{X^2} = z^m \left(\sum_j^{N/z} n_j^2 \right)^m + z \sum_j^{N/z} n_j^{2m} + z^2 g^2 \sum_{j \neq k}^{N/z} n_j^m n_k^m. \qquad (98)$$

The square of (90) may be written in the form

$$(\overline{X})^2 = z^2 g^2 \sum_{j,k}^{N/z} n_j^m n_k^m.$$

Consequently, the disperison

$$D = z^m \left\{ \sum_j^{N/z} n_j^2 \right\}^m + (z - z^2 g^2) \sum_j^{N/z} n_j^{2m}$$

or

$$D = \left\{ \sum_{j=1}^{N} n_j^2 \right\}^m + (1 - z g^2) \sum_{j=1}^{N} n_j^{2m}.$$

In the case of identical atoms

$$D = \frac{1}{N^m} + \frac{1 - z g^2}{N^{2m-1}} \approx \frac{1}{N^m}. \qquad (99)$$

Since the second term, as a rule, is negligibly small in comparison with the first one, we arrive at an important conclusion: The distribution function of the structure product for a more or less complex structure is not affected by the symmetry of the crystal.

The parameter D is also insensitive to zeros in the indices of the structure amplitudes.

12. The Probability of a Positive Sign for the Structure Product Within the Limits of Applicability of the Gaussian Distribution[8]

It is of great significance for the practice of structure analysis that the probability of a positive value of the structure product is greater than 1/2. This follows directly from the fact that the center of the distribution of X, according to (87), is shifted in the direction of positive values.

Let us find the probability W_+ that a structure product with the absolute magnitude $|X|$ is positive. The probability sought is equal to

$$W_+ = \frac{\varphi(X_+)}{\varphi(X_+) + \varphi(X_-)}\cdot$$

by definition. Substituting (87), we have

$$W_+ = \frac{1}{1 + e^{-2|X|\frac{\overline{X}}{D}}}\cdot \qquad (100)$$

The ratio $\frac{\overline{X}}{D}$ is the determinative parameter. We shall call

$$\frac{\overline{X}}{D} = N_{\text{eff}} \qquad (101)$$

the effective number of atoms. In the case of identical atoms in the cell $\frac{\overline{X}}{D} = N$, and (100) has the form

$$W_+ = \frac{1}{1 + e^{-2N|X|}}\cdot \qquad (102)$$

The greater the absolute magnitude $|X|$ the closer W_+ is to unity.

In order to gauge the possibility of applying (102), it is best to write it in the form

$$W_+ = \frac{1}{1 + e^{-2v\frac{\overline{X}}{\sqrt{D}}}}, \qquad (103)$$

[8] See page 116.

where $v = \frac{|X|}{\sqrt{D}}$ is a number which shows how many times X exceeds the magnitude of its root mean square $\left(D \approx \bar{X}^2, \text{ i.e., } v \approx \frac{|X|}{\sigma}\right)$. For identical atoms (103) has the form

$$W_+ = \frac{1}{1 + \exp\left(-2vN^{1-\frac{1}{2}m}\right)}. \tag{104}$$

From (104), it is evident that the chances of a certain determination of the positiveness of the structure product decrease as the number of atoms in the cell increases. For the same v, the closer the quantity W_+ is to unity, the smaller the number of atoms in the cell. Let us construct a table for structure products of the third and fourth orders. Besides the values of W_+, it is necessary to compute the integrated probabilities (fractions) of positive and negative structure products for large values of the moduli $|X|$.

Let us denote them thus:

$$P_+ = \frac{1}{\sqrt{2\pi D}} \int_X^{+\infty} e^{-\frac{(X-\bar{X})^2}{2D}} dX,$$

$$P_- = \frac{1}{\sqrt{2\pi D}} \int_{-X}^{-\infty} e^{-\frac{(X-\bar{X})^2}{2D}} dX.$$

Or, again introducing the variable $v = \frac{X}{\sqrt{D}}$

$$P_+ = \frac{1}{\sqrt{2\pi}} \int_{v-N^{\left(1-\frac{m}{2}\right)}}^{+\infty} e^{-\frac{1}{2}v^2} dv,$$

$$P_- = \frac{1}{\sqrt{2\pi}} \int_{-v-N^{\left(1-\frac{m}{2}\right)}}^{-\infty} e^{-\frac{1}{2}v^2} dv. \tag{105}$$

In computing these quantities one has to take account of the number of structure products which may be constructed for any one experiment. Assuming that the number of nodes of the reciprocal lattice whose weights are known is of the order of a thousand, the following may be accepted as the numbers of structure products:

m = 3	number	$X_{(3)}$	of the order $10^6 - 10^8$
m = 4	number	$X_{(4)}$	of the order $10^9 - 11^{11}$
m = 5	number	$X_{(5)}$	of the order $10^{12} - 10^{14}$ etc.

The results computed according to (104) and (105) are given in Table 4 for various N and m:

$$v = \frac{|X|}{\sqrt{\overline{X^2}}} \approx |X| N^{\frac{3}{2}}, \quad \overline{X} = \frac{1}{N^2}, \quad D = \frac{1}{N^3}, \quad \frac{\overline{X}}{\sqrt{D}} = \frac{1}{\sqrt{N}}.$$

Constructing tables for m > 4 obviously has no meaning. We come to the conclusion that only the triple (the simplest) product $\hat{F}_{\mathbf{H}}\hat{F}_{\mathbf{K}}\hat{F}_{\mathbf{H+K}}$ is of any practical significance for structure analysis.

Further discussion of the use in structure analysis of the preponderant positivity of the structure product is given in chapter IV.

13. Peculiarities of Incomplete Averaging of Structure Products

Up to now, it has been assumed that a complete (within the limits of the sphere of reflection) system of structure products was under study.

The problem is solved in another way if we analyze a group of structure products in which one of the indices h, k, l is constant.

Let us consider, for example, incomplete averaging for class C_{2h}. In all the groups of this class the structure amplitude for indices of different parity may be represented in one of the forms

$$\hat{F}_{U} = -\sum n_j \sin 2\pi (hx + lz) \sin 2\pi ky$$

or

$$\hat{F}'_{G} = \sum n_j \cos 2\pi (hx + lz) \cos 2\pi ky.$$

Let us call the odd amplitudes $\hat{F}_U$ and the even ones $\hat{F}'_G$.

Three types of structure products are possible: even–even, even – odd, and odd – odd.

Their mean values, over h and l with k and k' constant, are respectively equal to:

TABLE 4. Probabilities W_+, P_+, and P_- [9]

I. $m=3$

a) $N=25$; $\sqrt{D}=0{,}008$; $\overline{X}=\frac{1}{625}$

v	1	2	3	4	5	6	7
W_+	0.60	0.69	0.77	0.83	0.88	0.92	
P_+	0.211	0.036	$26 \cdot 10^{-4}$	$7.7 \cdot 10^{-5}$	$8.3 \cdot 10^{-7}$	$3.4 \cdot 10^{-9}$	
P_-	0.111	0.014	$6.9 \cdot 10^{-4}$	$1.4 \cdot 10^{-5}$	$1.0 \cdot 10^{-7}$	$2.9 \cdot 10^{-10}$	

b) $N=100$; $\sqrt{D}=0{,}001$; $\overline{X}=0{,}0001$

v	1	2	3	4	5	6	7
W_+	0.55	0.60	0.64	0.69	0.73	0.77	
P_+	0.18	0.029	$19 \cdot 10^{-4}$	$5.1 \cdot 10^{-5}$	$5.0 \cdot 10^{-7}$	$1.9 \cdot 10^{-9}$	
P_-	0.14	0.018	$9.7 \cdot 10^{-4}$	$2.2 \cdot 10^{-5}$	$1.8 \cdot 10^{-7}$	$5.5 \cdot 10^{-10}$	

II. $m=4$.

a) $N=25$

v	1	2	3	4	5	6	7
W_+	0.52	0.54	0.56	0.58	0.60	0.62	0.64
P_+	0,17	0,025	$15 \cdot 10^{-4}$	$4.0 \cdot 10^{-5}$	$3.7 \cdot 10^{-7}$	$1.3 \cdot 10^{-9}$	$1.8 \cdot 10^{-12}$
P_-	0,13	0,021	$11 \cdot 10^{-4}$	$2.8 \cdot 10^{-5}$	$2.4 \cdot 10^{-7}$	$8.0 \cdot 10^{-10}$	$10 \cdot 10^{-13}$

b) $N=100$

v	1	2	3	4	5	6	7
W_+	0.50	0.51	0.51	0.52	0.52	0.53	0.53
P_+	0,14	0,023	$14 \cdot 10^{-4}$	$3.5 \cdot 10^{-5}$	$3.2 \cdot 10^{-7}$	$1.1 \cdot 10^{-9}$	$1.4 \cdot 10^{-12}$
P_-	0,13	0,022	$13 \cdot 10^{-4}$	$3.2 \cdot 10^{-5}$	$2.8 \cdot 10^{-7}$	$9.6 \cdot 10^{-10}$	$1.2 \cdot 10^{-12}$

[9] The values of the probability integral for large arguments are computed according to the formula $1+\Phi(x) \approx \frac{e^{\frac{1}{2}x^2}}{\sqrt{2\pi}}$.

$$\bar{X}_{G,G} = \overline{\cos 2\pi(hx_j + lz_j)\cos 2\pi(h'x_j + l'z_j)\cos 2\pi[(h+h')x_j + (l+l')z_j]} \times$$
$$\times \sum n_j^3 \cos 2\pi ky_j \cos 2\pi k'y_j \cos 2\pi(k+k')x_j;$$

$$\bar{X}_{U,G} = \overline{\sin 2\pi(hx_j + lz_j)\cos 2\pi(h'x_j + l'z_j)\sin 2\pi[(h+h')x_j + (l+l')z_j]} \times$$
$$\times \sum n_j^3 \sin 2\pi ky_j \cos 2\pi k'y_j \sin 2\pi(k+k')y_j;$$

$$\bar{X}_{U,U} = \overline{\sin 2\pi(hx_j + lz_j)\sin 2\pi(h'x + l'z_j)\cos 2\pi[(h+h')x_j + (l+l')z_j]} \times$$
$$\times \sum n_j^3 \sin 2\pi ky_j \sin 2\pi k'y_j \cos 2\pi(k+k')y_j.$$

The mean values, standing in front of the summation signs are respectively equal to $+\frac{1}{4}$, $+\frac{1}{4}$ and $-\frac{1}{4}$.

$$\bar{X}_{G,G} = +\frac{1}{4}\sum n_j^3 \cos 2\pi ky_j \cos 2\pi k'y_j \cos 2\pi(k+k')y_j;$$
$$\bar{X}_{U,G} = +\frac{1}{4}\sum n_j^3 \sin 2\pi ky_j \cos 2\pi k'y_j \sin 2\pi(k+k')y_j;$$
$$\bar{X}_{U,U} = -\frac{1}{4}\sum n_j^3 \sin 2\pi ky_j \sin 2\pi k'y_j \cos 2\pi(k+k')y_j.$$

It is obvious from these expressions that $\bar{X}$ may be, generally speaking, either positive or negative. Let us note that when $k = 0$ all the $\bar{X} > 0$.

If there is one heavy atom, the sign of $\bar{X}$ will be determined only by its y coordinates (the factor n_j^3).

When combining the reflections $h1l$ and $h'1l'$, the following interesting relationships arise, for example; $\bar{X}_{UG}$ is always > 0, $\bar{X}_{GG} > 0$ and $\bar{X}_{UU} < 0$, if $\left|\sum n_j^3 \cos 2y_j\right| > \sum n_j^3 \cos^2 2y_j$ and $\sum n_j^3 \cos 2y_j > 0$. The inverse relationships will exist when $\sum n_j^3 \cos 2y_j < 0$.

Let us note that this result, which is derived from the general rule, is always obtained in the case of a heavy atom (since cos 2y is always larger than $\cos^2 2y$), simultaneously $X_{GG} > 0$ and $X_{UU} > 0$ only in the case that $|\sum n_j^3 \cos 2y_j| < \sum n_j^3 \cos^2 2y_j$.

Thus, with incomplete averaging it is possible that some groups of structure products have a greater probability of being negative.

It is clear from the example given, that the effect indicated is absent in triclinic crystals; in monoclinic ones it exists for groups H and K with constant k and k' or h, l and h',l'; in orthorhomic crystals peculiarities occur in the behavior of groups of structure products with identical h,h', and also k, k' and l,l' The same effect shows up clearly for structure products with identical l,l' in the classes D_{4h} and D_{6h}, etc.

CHAPTER IV

THE THEORY OF THE RELATIONS BETWEEN STRUCTURE AMPLITUDES

1. Statement of the Problem

The experimentally measured intensities of diffracted rays make it possible to compute directly only the moduli of the structure amplitudes. At first glance, the problem of determining the phases of the structure amplitudes by means of the experimental intensity data alone may appear impossible.

However, this is not so. The number of the experimentally measured moduli of structure amplitudes for reflections of different orders is of the order of many hundreds and even thousands. The values of all these moduli are determined by a relatively small number of atomic coordinates. For each node of the reciprocal lattice, (1) holds:

$$|F_{\mathbf{H}}| = \left| \sum_{j=1}^{N} f_j \exp i\alpha_j \right|, \tag{1}$$

where

$$\alpha_j = 2\pi \mathbf{H} \mathbf{r}_j = 2\pi (hx_j + ky_j + lz_j). \tag{2}$$

The number of these equations exceeds usually by one or two orders the number of unknowns 3N. It is clear that in principle a structure determination is possible by solving a system of 3N equations, or, if the possibility of a many-valued solution is taken into account , by solving a number of simultaneous equations exceeding 3 N, in any case, by no more than an order of magnitude.

Hence, it follows that there must be connections between the structure amplitudes (both between their moduli and phases). The necessity for connection between the structure amplitudes is also due to the fact that the electron density of a crystal is a nonnegative function. For every value of xyz the Fourier series

$$\sum_{\mathbf{H}} F_{\mathbf{H}} \exp - 2\pi i (hx + ky + lz) \geqslant 0. \tag{3}$$

This relationship imposes conditions on the values of the moduli and the phases of the structure amplitudes.

The basic task of this chapter is the study of reliable connections between the structure amplitudes. Our main goal is the finding of relationships on the basis of which it may be possible to make judgements concerning the phases of the structure amplitudes knowing their moduli. "A reliable connection" is juxtaposed to the probable connections, which were discussed in the preceding chapter, and to which we shall return at the end of this one.

In spite of the fact that in principle the problem should be solved for a crystal of any symmetry, we shall emphasize here the study only of such crystals as have a center of symmetry. This is a great simplification, since, instead of determining the magnitude of the phase angle, only the sign of the structure amplitude has to be found.

Articles have recently appeared in the literature devoted to the utilization of connections between structure amplitudes for the purpose of finding the phase angles of the amplitudes of reflections for crystals without centers of symmetry. However, it is too soon to discuss this problem.

We shall see that for a crystal with a center of symmetry, in practice (in a reasonable time) the feasible determination of the signs of structure amplitudes from their moduli with the aid of reliable connections between moduli becomes less possible as the number of atoms in the unit cell of the crystal increases. However, even small success is important.

As we have seen in Chapter III, statistical considerations allow one to determine that a structure product has a positive sign with a probability greater than 1/2. If some of the signs of the structure amplitudes are known, then the use of the formulas in Chapter III, section 11 gives us a method for determining all the signs of the structure amplitudes. For this it is necessary only to construct a large number of structure products in which one and the same amplitude occurs many times.

To carry out the statistical method of determining the signs reliably, it is necessary to know with certainty at least a small number of signs. Sign determination is a "chain process." Therefore a mistake in the initial stage of the analysis will entail a series of errors.

Hence, the importance of reliable connections between structure amplitudes that create a basis for the study of the signs by the statistical method. Statistical theory has a limited applicability — its formulas are true only for not too large structure products.

The theory of the connections between structure amplitudes discussed in this chapter makes possible the finding of relationships between amplitudes exactly in that region in which the statistical theory fails. The consummation of the theory is the synthesis of the reliable connections between structure amplitudes with the connections established by the statistical theory. In Section 20, the complete solution of this problem is given.

2. Reliably Positive Structure Products [21]

First of all, let us show that in the simplest structures the signs of the amplitudes are easily determined by approximating the absolute magnitude of structure products. The outstanding characteristic of structure products is formulated in the following theorem. If $|X_{HKL\ldots(H+K+L\ldots)}| > \delta$, then the structure product is reliably positive. The magnitude δ depends on the order m of the structure product, that is,

$$\delta = \cos^{m-1}\left(\frac{180^\circ}{m}\right)\cos(m-1)\frac{180^\circ}{m}. \tag{4}$$

The values of δ are given in the table

m	3	4	5	6
reliably positive X larger than	0.125	0.250	0.347	0.423

and so forth.

Let us prove (4). The ultimate positive value of a structure product is equal to unity. But the minimum negative value is less than unity and is equal to δ. For a proof one has to find the minimum value of X for any values of the arguments of the trigonometric functions α_j, β_j and so forth, i.e., the derivatives must be computed in the form

$$\frac{\partial X}{\partial \alpha_k} = \frac{\partial}{\partial \alpha_k}\left\{2^m \sum n_j \cos\alpha_j \sum n_j \cos\beta_j \ldots \sum n_j \cos(\alpha_j + \beta_j + \ldots)\right\}$$

and equated to zero. The condition of the extreme has the same form for all α_j, β_j, etc., namely,

$$\frac{\sin\alpha}{\hat{F}_{\mathbf{H}}} = -\frac{\sin(\alpha+\beta+\ldots)}{\hat{F}_{\mathbf{H+K+\ldots}}}. \tag{5}$$

The minimum of the structure factor will arise when all the arguments α_j, β_j and so forth, are equal to one another. Substituting in (5) $\hat{F}_{\mathbf{H}} = \cos\alpha$ and $\hat{F}_{\mathbf{H+K+\ldots}} = \cos(\alpha+\beta+\ldots)$, we obtain the condition for the minimum in the form $\tan\alpha = -\tan(m-1)\alpha$, which gives $\alpha = \frac{180°}{m}$ and, consequently, for X (4).

Structure products of the third order are of practical significance. We have proved that $-\frac{1}{8} < \hat{F}_{\mathbf{H}}\hat{F}_{\mathbf{K}}\hat{F}_{\mathbf{H+K}} < 1$. Thus, under the condition that

$$|\hat{F}_{\mathbf{H}}\hat{F}_{\mathbf{K}}\hat{F}_{\mathbf{H+K}}| > \frac{1}{8} \tag{6}$$

the structure product is reliably positive.

Let us note, that less deep minima in the product of the third order will arise when there is a linear relationship between **H** and **K**. Let us assume that $\mathbf{K} = \epsilon\mathbf{H}$. Repeating the computations of the derivatives, we obtain the condition of the extreme in the form

$$\mathrm{tg}\,\alpha + \varepsilon\,\mathrm{tg}\,\varepsilon\alpha + (1+\varepsilon)\,\mathrm{tg}\,(1+\varepsilon)\,\alpha = 0.$$

The deepest minima are as follows:

ε	1	$^5/_4$	$^3/_2$	2	3	4
$X_{\min}$	—0.125	—0.118	—0.106	—0.079	—0.092	—0.125

The table shows that $X_{\min} > -\frac{1}{8}$ will occur in the triclinic system only for indices with two zeros, in the monoclinic system only for indices of the two-fold axis zone and in the orthorhombic and other systems for indices of the general type.

It is obvious that (6) will be useful only in case there is at least a small number of $|\hat{F}| > \frac{1}{2}$ among the structure amplitudes. If this is fulfilled, then (6) is of course the simplest method for sign determination that exists.

The possibilities of (6) are discussed in Chapter IV, Section 18. However, we note at once that ten structure amplitudes greater than 1/2, will give 50-100 reliably positive products, which is quite sufficient for the initial stage of sign determination by the statistical method.

Such will be the case for structures with a number up to 30-40, of identical atoms in the cell.

3. The Centrosymmetric Crystal with One Atom in a General Position

We analyze this simplest case only to facilitate the understanding of the genesis of the general relationships. For a centrosymmetric crystal with two atoms in the cell, the unitary structure amplitude is equal to

$$\hat{F}_{\mathbf{H}} = \cos \alpha_{\mathbf{H}},$$

where $\alpha_{\mathbf{H}} = 2\pi \mathbf{Hr}$.

All possible connections between structure amplitudes arise from the possibility of constructing trigonometric formulas that connect cosines.

Among these, for example, are the following:

$$2\cos^2\alpha = 1 + \cos 2\alpha; \tag{7}$$

$$\cos 3\alpha = 4\cos^3\alpha - 3\cos\alpha; \tag{8}$$

$$2\cos\alpha\cos\beta = \cos(\alpha+\beta) + \cos(\alpha-\beta); \tag{9}$$

$$\cos^2\alpha + \cos^2\beta = 1 + \cos(\alpha+\beta)\cos(\alpha-\beta); \tag{10a}$$

$$(\cos\alpha + \cos\beta)^2 = 1 + \cos(\alpha+\beta)\cos(\alpha-\beta) + \cos(\alpha-\beta) + \cos(\alpha+\beta). \tag{10b}$$

From these arise, respectively, such equations as

$$\left.\begin{aligned} &2\hat{F}^2_{\mathbf{H}} = 1 + \hat{F}_{2\mathbf{H}};\\ &\hat{F}_{3\mathbf{H}} = 4\hat{F}^3_{\mathbf{H}} - 3\hat{F}_{\mathbf{H}};\\ &2\hat{F}_{\mathbf{H}}\hat{F}_{\mathbf{K}} = \hat{F}_{\mathbf{H+K}} + \hat{F}_{\mathbf{H-K}};\\ &\hat{F}^2_{\mathbf{H}} + \hat{F}^2_{\mathbf{K}} = 1 + \hat{F}_{\mathbf{H+K}}\hat{F}_{\mathbf{H-K}};\\ &\hat{F}^2_{\mathbf{II}} + \hat{F}^2_{\mathbf{K}} + \hat{F}^2_{\mathbf{H-K}} - 1 = 2\hat{F}_{\mathbf{H}}\hat{F}_{\mathbf{K}}\hat{F}_{\mathbf{H-K}}. \end{aligned}\right\} \tag{11}$$

It is quite obvious that similar equations solve the structure problem exhaustively, since knowing the absolute values of the quantities $\hat{F}$, it is possible to find their signs.

However, these equations are correct only for the case of one independent (and approximate for the case of one heavy) atom in the cell. Therefore, their value is negligible.

Of interest is the construction of similar equations for a crystal in whose cell there is any number of atoms.

4. Relationships Between Amplitudes and Their Components

For the general case of a centrosymmetric crystal, relationships analogous to those just written out are fulfilled, not for the structure amplitudes themselves, but for their components.

If we multiply (7)-(10) or other similar equations by n_j and sum over the independent atoms, we may obtain identities of the following type:

$$\left.\begin{aligned} 2\sum_{1}^{N/2} n_j \cos^2 2\pi \mathbf{H}\mathbf{r}_j &= \frac{1}{2} + \frac{1}{2}\hat{F}_{2\mathbf{H}}; \\ \hat{F}_{3\mathbf{H}} &= 4\sum n_j \cos^3 2\pi\mathbf{H}\mathbf{r}_j - 3\hat{F}_{\mathbf{H}}; \\ 2\sum n_j \cos 2\pi\mathbf{H}\mathbf{r}_j \cos 2\pi\mathbf{K}\mathbf{r}_j &= \hat{F}_{\mathbf{H}+\mathbf{K}} + \hat{F}_{\mathbf{H}-\mathbf{K}} \end{aligned}\right\} \qquad (12)$$

and so forth.

It is possible to do likewise for the relationships in (10), writing them out in somewhat different form. However, one should not linger on this. As a matter of fact, the derivation of such relationships [35] comes out the same with the use of the identity

$$|\sum a_j b_j|^2 = \sum |a_j|^2 \sum |b_j|^2 - \frac{1}{2}\sum_{j \neq m}\sum \begin{vmatrix} a_j a_m \\ b_j b_m \end{vmatrix}^2. \qquad (13)$$

Assuming, for example that $a_j = (2n_j)^{\frac{1}{2}}$ and $b_j = (2n_j)^{\frac{1}{2}} \cos 2\pi\mathbf{H}\mathbf{r}_j$, we obtain

$$\left|\sum_{j=1}^{N/2} a_j b_j\right|^2 = \hat{F}^2_{\mathbf{H}},$$

$$\sum_{j=1}^{N/2} |a_j|^2 = 1,$$

$$\sum_{j=1}^{N/2} |b_j|^2 = 2\sum_{j=1}^{N/2} n_j \cos^2 2\pi\mathbf{H}\mathbf{r} = \frac{1}{2} + \frac{1}{2}\hat{F}_{2\mathbf{H}}.$$

Consequently,

$$\hat{F}^2_{\mathbf{H}} = \frac{1}{2} + \frac{1}{2}\hat{F}_{2\mathbf{H}} - 2\sum_{j \neq m}^{N/2}\sum^{N/2} n_j n_m |\cos 2\pi\mathbf{H}\mathbf{r}_m - \cos 2\pi\mathbf{H}\mathbf{r}_j|. \qquad (14)$$

Let us consider one more special case of (13), i.e., let us assume

$$a_j=(2n_j)^{\frac{1}{2}}\cos\frac{1}{2}2\pi(\mathbf{H}+\mathbf{K})\mathbf{r}_j \quad \text{и} \quad b_j=(2n_j)^{\frac{1}{2}}\cos\frac{1}{2}2\pi(\mathbf{H}-\mathbf{K})\mathbf{r}_j;$$

$$\left|\sum_{j=1}^{N/2}a_jb_j\right|^2=\left|2\sum_{j=1}^{N/2}n_j\cos\frac{1}{2}2\pi(\mathbf{H}+\mathbf{K})\mathbf{r}_j\cos\frac{1}{2}2\pi(\mathbf{H}-\mathbf{K})\mathbf{r}_j\right|^2=$$

$$=\left|\sum_{1}^{N/2}n_j\cos 2\pi\mathbf{H}\mathbf{r}_j+\sum_{1}^{N/2}n_j\cos 2\pi\mathbf{K}\mathbf{r}_j\right|^2=\frac{1}{4}|\hat{F}_{\mathbf{H}}+\hat{F}_{\mathbf{K}}|^2;$$

$$\sum_{1}^{N/2}|a_j|^2=2\sum_{1}^{N/2}n_j\cos^2 2\pi\frac{1}{2}(\mathbf{H}+\mathbf{K})\mathbf{r}_j=\frac{1}{4}+\frac{1}{4}\hat{F}_{\mathbf{H}+\mathbf{K}};$$

$$\sum_{1}^{N/2}|b_j|^2=\frac{1}{4}(1+\hat{F}_{\mathbf{H}-\mathbf{K}}).$$

Consequently,

$$|\hat{F}_{\mathbf{H}}+\hat{F}_{\mathbf{K}}|^2=1+\hat{F}_{\mathbf{H}+\mathbf{K}}+\hat{F}_{\mathbf{H}-\mathbf{K}}+\hat{F}_{\mathbf{H}-\mathbf{K}}\hat{F}_{\mathbf{H}+\mathbf{K}}-D_{\mathbf{HK}}, \tag{15}$$

where

$$D_{\mathbf{HK}}=8\sum_{j\neq m}^{N/2}\sum^{N/2}n_jn_m\left|\cos\frac{1}{2}2\pi(\mathbf{H}+\mathbf{K})\mathbf{r}_j\cos\frac{1}{2}2\pi(\mathbf{H}-\mathbf{K})\mathbf{r}_m-\right.$$
$$\left.-\cos\frac{1}{2}2\pi(\mathbf{H}+\mathbf{K})\mathbf{r}_m\cos\frac{1}{2}2\pi(\mathbf{H}-\mathbf{K})\mathbf{r}_j\right|^2. \tag{16}$$

Assuming

$$a_j=(2n_j)^{\frac{1}{2}}\sin\frac{1}{2}2\pi(\mathbf{H}+\mathbf{K})\mathbf{r}_j \quad \text{и} \quad b_j=(2n_j)^{\frac{1}{2}}\sin\frac{1}{2}2\pi(\mathbf{H}-\mathbf{K})\mathbf{r}_j,$$

we obtain the relationship,

$$|F_{\mathbf{H}}-F_{\mathbf{K}}|^2=1-F_{\mathbf{H}+\mathbf{K}}-F_{\mathbf{H}-\mathbf{K}}+F_{\mathbf{H}+\mathbf{K}}F_{\mathbf{H}-\mathbf{K}}-G_{\mathbf{HK}}, \tag{17}$$

where

$$G_{\mathbf{HK}}=8\sum\sum n_jn_m\left|\sin\frac{1}{2}2\pi(\mathbf{H}+\mathbf{K})\mathbf{r}_j\sin\frac{1}{2}2\pi(\mathbf{H}-\mathbf{K})\mathbf{r}_m-\right.$$
$$\left.-\sin\frac{1}{2}2\pi(\mathbf{H}+\mathbf{K})\mathbf{r}_m\sin\frac{1}{2}2\pi(\mathbf{H}+\mathbf{K})\mathbf{r}_j\right|^2. \tag{18}$$

We have given two simple methods for finding correlations between structure amplitudes and their components. Since the components enter these equations, it is difficult to render them practicable.

However, some useful consequences may be drawn from such equations. They lead to inequalities relating structure amplitudes. This will be discussed below.

5. Concerning the Averaging of the Relationships Between Unitary Structure Amplitudes and Their Components

Such averaging may be carried out only if we assume that when averaging over one or several indices, the arguments of the trigonometric quantities into which this index enters are distributed uniformly over the trigonometric circle. The correctness of such an assumption has been discussed in Chapter III.

Such averaging leads, for example, to the following results:

$$\overline{\hat{F}_{\mathbf{H}}}^{\mathbf{H}} = \overline{\hat{F}_{\mathbf{H+K}}}^{\mathbf{H}} = 0, \tag{19}$$

$$\overline{\hat{F}^2_{\mathbf{H}}} = 2\sum_{j=1}^{N/2} n_j^2; \tag{20}$$

$$\overline{\hat{F}_{\mathbf{H+K}}\hat{F}_{\mathbf{H}}}^{\mathbf{H}} = 2\sum_{j=1}^{N/2} n_j^2 \cos 2\pi\mathbf{K}\mathbf{r}_j \tag{21}$$

etc., which may be easily checked by substituting $\hat{F}_{\mathbf{H}} = 2\sum_{j=1}^{N/2} n_j \cos 2\pi\mathbf{H}\mathbf{r}_j$ and averaging, on the assumption that

$$\overline{\cos 2\pi\mathbf{H}\mathbf{r}_j} = 0 \quad \text{and} \quad \overline{\cos^2 2\pi\mathbf{H}\mathbf{r}_j} = \frac{1}{2}.$$

When there is a sufficiently large number of terms used in the averaging and, of course, when there is no selection of terms which might break the condition of equal distribution of the argument,[1] (19)-(21) are fulfilled with a sufficient accuracy which increases with the averaging volume.

[1]The only accurate averaging in this sense is averaging over spherical shells of reciprocal space. In this case a variation of the argument is guaranteed between the limits from zero to whole multiples of π. One has to keep this in mind, especially for (21): the node **K** is arbitrary, the nodes **H** fill the spherical region.

Equation (21) acquires an especially simple form for identical atoms. Obviously,

$$\overline{\hat{F}_{\mathrm{H+K}}\hat{F}_{\mathrm{H}}}^{\mathrm{H}}=\frac{1}{N}\hat{F}_{\mathrm{K}}. \tag{22}$$

Practical application can be made of this correlation. Equation (22) possibly also has some meaning for nonunitary amplitudes.

Equation (22), which for the first time (it is true, in somewhat different form) was presented by Sayre [36] is of historical significance. The appearance of this equality has led other researchers to the idea of a probable connection between the signs of $\hat{F}_{\mathbf{K}}$ and $\hat{F}_{\mathbf{H+K}}\hat{F}_{\mathbf{H}}$ [although, of course, such a connection does not follow directly from (22)].

The averaging of relations between unitary structure amplitudes and their components may be carried out only under the condition mentioned at the beginning of this section. Therefore, it will always lead us to equations of the type (19)-(21), i.e., to trivial results.

Let us pause on the question of averaging the relations analyzed in the preceding paragraph. This averaging has been discussed in the literature many times and has led to erroneous conclusions ([35], [37], and [38]).

Let us carry out in (15) a substitution of indices and write it in the form

$$|\hat{F}_{\mathrm{H+K}}+\hat{F}_{\mathrm{K}}|^2=1+\hat{F}_{\mathrm{H}}+\hat{F}_{\mathrm{H+2K}}+\hat{F}_{\mathrm{H}}\hat{F}_{\mathrm{H+2K}}-D_{\mathrm{H+K,\,K}}, \tag{23}$$

where

$$D_{\mathrm{H+K,\,K}}=8\sum_{j\neq m}^{N/2}\sum^{N/2}n_j n_m\left|\cos\frac{1}{2}2\pi\mathbf{H}\mathbf{r}_j\cos\frac{1}{2}2\pi(\mathbf{H}+2\mathbf{K})\mathbf{r}_m-\right.$$
$$\left.-\cos\frac{1}{2}2\pi\mathbf{H}\mathbf{r}_m\cos\frac{1}{2}2\pi(\mathbf{H}+2\mathbf{K})\mathbf{r}_j\right|^2; \tag{24}$$

and let us carry out an averaging over all the nodes **K**.

Squaring the trigonometric term in (24), applying the formula of the double angle, and carrying out the averaging over **K** on the assumption that the values of **Kr** are equally distributed over the trigonometric circle, we find

$$\overline{D_{\mathrm{H+K,K}}}=1+\hat{F}_{\mathrm{H}}-4\sum_{1}^{N/2}n_j^2(1+\cos 2\pi\mathbf{H}\mathbf{r}_j),$$

substituting in (23), and taking into account that $\bar{\hat{F}}_{\mathbf{K}} = 0$, $\bar{\hat{F}}_{\mathbf{H}+2\mathbf{K}} = 0$, we have

$$\overline{(\hat{F}_{\mathbf{H}+\mathbf{K}}+\hat{F}_{\mathbf{K}})^2}=4\sum_{1}^{N/2} n_j^2(1+\cos 2\pi\mathbf{H}\mathbf{r}_j). \tag{25}$$

Carrying out a similar conversion for (17) we obtain

$$\left.\begin{aligned}\overline{G_{\mathbf{H}+\mathbf{K},\,\mathbf{K}}}&=1-\hat{F}_{\mathbf{H}}-4\sum_{1}^{N/2} n_j^2(1-\cos 2\pi\mathbf{H}\mathbf{r}_j);\\ \overline{(\hat{F}_{\mathbf{H}+\mathbf{K}}-\hat{F}_{\mathbf{K}})^2}&=4\sum_{1}^{N/2} n_j^2(1-\cos 2\pi\mathbf{H}\mathbf{r}_j).\end{aligned}\right\} \tag{26}$$

Adding (25) and (26), we arrive at a trivial result:

$$\overline{\hat{F}^2_{\mathbf{H}+\mathbf{K}}}+\overline{\hat{F}^2_{\mathbf{K}}}=4\sum_{j=1}^{N/2} n_j^2,$$

subtracting, we arrive at another result:

$$\overline{\hat{F}_{\mathbf{H}+\mathbf{K}}\hat{F}^{\mathbf{K}}_{\mathbf{K}}}=2\sum_{j=1}^{N/2} n_j^2\cos 2\pi\mathbf{H}\mathbf{r}_j.$$

It is possible to assume that averaging other analogous relations will not lead to anything new.

6. The Simplest Inequality Relating Unitary Structure Amplitudes

Thus, relations between structure amplitudes and their components have in themselves little practical application in structure analysis. From these relations, however, connections between unitary structure amplitudes emerge graphically in the form of inequalities.

The derivation of the inequalities is most simply realized with the aid of Cauchy's relation, which derives from the inequality given in Section 4 (13). Since the second term of this inequality is essentially positive, then

$$\left|\sum a_j b_j\right|^2 \leqslant \sum a_j^2 \sum b_j^2. \tag{27}$$

Formula (27) expresses, as is known, in the language of multidimensional geometry, the fact that the scaler product is smaller than the product of the vector lengths.

Let us illustrate the application of this inequality to our goals by the simplest example. Let us assume that $a_j = n_j^{\frac{1}{2}}$ and $b_j = n_j^{\frac{1}{2}} \cos\alpha_j$. Then

$$\left|\sum_1^N a_j b_j\right|^2 = \left|\sum_1^N n_j \cos\alpha_j\right|^2 = \hat{F}^2_{\mathbf{H}},\quad \sum_1^N a_j^2 = 1 \quad \text{и} \quad \sum_1^N b_j^2 = \sum_1^N n_j \cos^2\alpha_j.$$

Thus, from (27) it follows that

$$\hat{F}^2_{\mathbf{H}} \leqslant \sum_1^N n_j \cos^2\alpha_j.$$

But, if this is so, then the first equation in (12) gives

$$\hat{F}^2_{\mathbf{H}} \leqslant \frac{1}{2} + \frac{1}{2}\hat{F}_{2\mathbf{H}}. \tag{28}$$

This simplest inequality is derived in Harker and Kasper's [11] article, which was the impetus for the study of the connection between structure amplitudes. Inequality (28), like all the inequalities which we shall study below, has its own region of application in the field of the absolute magnitudes $|\hat{F}_{2\mathbf{H}}|$ and $|\hat{F}_{\mathbf{H}}|$. The inequality is applicable only when some choice of sign contradicts it. For example, if $|\hat{F}_{\mathbf{H}}| = 0.7$, but $|\hat{F}_{2\mathbf{H}}| = 0.5$, then $\hat{F}_{2\mathbf{H}}$ cannot be negative, since the left side of the inequality is equal to 0.49, and the right side is equal to 0.25.

Let us examine the absolute values of the structure amplitudes $|\hat{F}_{\mathbf{H}}|$ and $|\hat{F}_{2\mathbf{H}}|$. The magnitude $\hat{F}^2_{\mathbf{H}}$ must be smaller than $\frac{1}{2} + \frac{1}{2}\hat{F}_{2\mathbf{H}}$. Depending on the sign of $\hat{F}_{2\mathbf{H}}$ the last expression is either $\frac{1}{2} + \frac{1}{2}|\hat{F}_{2\mathbf{H}}|$ or $\frac{1}{2} - \frac{1}{2}|\hat{F}_{2\mathbf{H}}|$. From (28) it follows that: 1) $\hat{F}^2_{\mathbf{H}}$ cannot be greater than $\frac{1}{2} + \frac{1}{2}|\hat{F}_{2\mathbf{H}}|$, and consequently, must be in every case smaller than $\frac{1}{2} + \frac{1}{2}|\hat{F}_{2\mathbf{H}}|$; 2) if $\hat{F}^0_{\mathbf{H}} < \frac{1}{2} - \frac{1}{2}\hat{F}_{2\mathbf{H}}$, then inequality (28) does not determine a sign.

From which it follows that inequality (28) leads to a sign determination, if

$$\frac{1}{2} - \frac{1}{2}|\hat{F}_{2\mathbf{H}}| < \hat{F}^2_{\mathbf{H}} < \frac{1}{2} + \frac{1}{2}|\hat{F}_{2\mathbf{H}}|.$$

What has been said has a simple geometric meaning, illustrated in Fig. 12.

Two parabolas $y^2 = \frac{1}{2} - \frac{1}{2}\,|x|$ and $y^2 = \frac{1}{2} + \frac{1}{2}\,|x|$ limit the field of action of inequaltiy (28). In Fig. 12 the region of impossible values is shaded; 0 marks the region which does not give a sign determination.

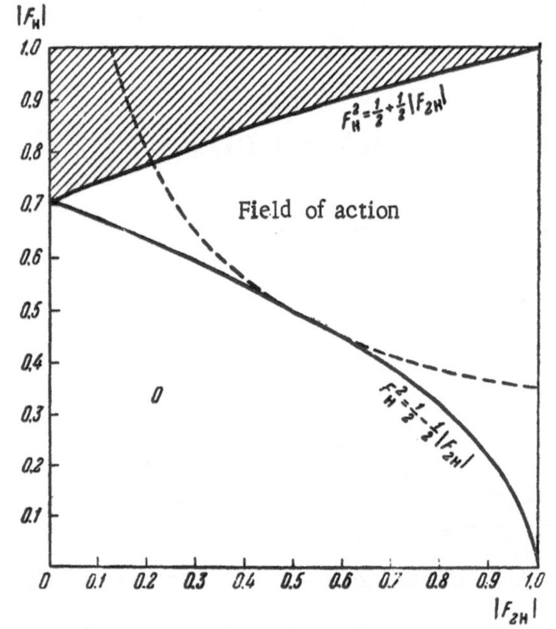

Fig. 12. The field of action of the simplest Harker-Kasper inequality.

It follows from the graph that when $|\hat{F}_H| = 0.7$, the field of action lies at $|\hat{F}_{2H}| > 0.02$; if $|\hat{F}_H| = 0.6$ then at $|\hat{F}_{2H}| > 0.28$; if $|\hat{F}_H| = 0.5$, then at $|\hat{F}_{2H}| > 0.5$.

Unfortunately, the experimental points for a complex structure fall frequently into the region of zero result[2] which makes a practical result from this simplest inequality hardly probable.

[2] As we have seen (Chapter III) the magnitudes of $\hat{F}_H$ obey the Gaussian distribution with a dispersion $\frac{1}{\sqrt{N}}$ for a structure of N identical atoms. Therefore large values of $\hat{F}_{2H}$ and $\hat{F}_H$ occur rarely.

It is best to judge the possibility of using an inequality by its field of action; however, it is not always a simple problem to construct this region. On the other hand, as we shall see below, it is easy to determine a fraction ϵ which satisfies the following condition. If all the $|\hat{F}|$ that enter the inequality are smaller than ϵ, then the inequality does not determine a sign. We shall call the number ϵ the limit of applicability of the inequality, or briefly, the boundary. The boundary of (28) is obviously equal to $\epsilon = \frac{1}{2}$ (the root of the equation $x^2 = \frac{1}{2} - \frac{1}{2}x$).

Let us point out one characteristic of (28): it either determines a positive sign for $\hat{F}_{2H}$ or does not determine any sign at all. This is not a property common to all inequalities. However, for the inequality at hand this circumstance has a general basis (see page 179). Inequalities determine the signs of structure products.

It is interesting to compare the possibility of using inequality (28) with the conditions in (6). According to (6) $F_{2H} > 0$, if $|\hat{F}_{2H}| > \frac{1}{8\hat{F}_H^2}$. The curve $y^2 = \frac{1}{8x}$ is plotted in the field of action of inequality (28). We see that the possibility of determining a sign by condition (6) differs negligibly from that of inequality (28). The boundaries ϵ of both conditions completely coincide in the middle part of the graph.

The considerations discussed above will be reexamined in a general way in Section 19.

7. Derivation of the Relationships Between Structure Amplitudes Using Cauchy's Inequality. The First Method

With the aid of Cauchy's inequality (27), Harker and Kasper obtained a series of useful correlations between structure amplitudes. It will be shown below that the use of Cauchy's inequality is not a general method for this purpose.

The first method consists in equating the left side of inequality (27) to a unitary structure amplitude,

$$\sum_{j=1}^{N} a_j b_j = \hat{F}_{hkl}.$$

Depending on the expressions $\hat{F}_{hkl}$ in the various space groups, the breaking of the components of the structure amplitudes into factors a_j and b_j can be performed in different ways.

For a crystal devoid of symmetry elements, we have

$$\sum_{j=1}^{N} a_j b_j = \sum_{j=1}^{N} n_j \exp 2\pi i H r_j,$$

i.e., $a_j b_j = n_j \exp 2\pi i H r_j$.

Any breaking into factors, followed by a subsequent substitution in (27), will lead only to the trivial result $\hat{F}^2 \le 1$.

It is otherwise for a symmetrical crystal. Let us analyze the groups which contain only one element of symmetry.

Group $\bar{1}$

$$\sum_{j=1}^{N} a_j b_j = 2 \sum_{j=1}^{N/2} n_j \cos 2\pi \mathrm{Hr}_j.$$

Assuming

$$a_j = (2n_j)^{\frac{1}{2}}; \quad b_j = (2n_j)^{\frac{1}{2}} \cos 2\pi \mathrm{Hr}_j$$

and substituting in (27) we obtain

$$\hat{F}_{\mathrm{H}}^2 \leqslant 2 \sum_{j=1}^{N/2} n_j 2 \sum_{j=1}^{N/2} n_j (\cos 2\pi \mathrm{Hr}_j)^2,$$

or, recalling that $\cos^2 \alpha = \frac{1}{2}(1 + \cos 2\alpha)$ and $\sum_{j=1}^{N/2} n_j = \frac{1}{2}$:

$$\hat{F}_{\mathrm{H}}^2 \leqslant \sum_{j=1}^{N/2} n_j (1 + \cos 2\pi \mathrm{Hr}_j),$$

or

$$\hat{F}_{\mathrm{H}}^2 \leqslant \frac{1}{2} + \frac{1}{2} \hat{F}_{2\mathrm{H}}.$$

We have repeated the derivation of (28).

Group P2

$$\sum a_j b_j = 2 \sum_{j=1}^{N/2} n_j e^{2\pi i k y_j} \cos 2\pi (h x_j + l z_j).$$

Let us assume

$$a_j = (2n_j)^{\frac{1}{2}} e^{2\pi i k y_j} \text{ and } b_j = (2n_j)^{\frac{1}{2}} \cos 2\pi (h x_j + l z_j).$$

The inequality acquires the form

$$\hat{F}_{\mathbf{H}}^2 \leqslant 4 \sum_{j=1}^{N/2} n_j \left| e^{2\pi i k y_j} \right|^2 \sum_{j=1}^{N/2} n_j \cos^2 2\pi (h x_j + l z_j),$$

or

$$\hat{F}_{\mathbf{H}}^2 \leqslant 4 \sum_{j=1}^{N/2} n_j \sum_{j=1}^{N/2} n_j \cos^2 2\pi (h x_j + l z_j),$$

which gives

$$\hat{F}_{hkl}^2 \leqslant \frac{1}{2} + \frac{1}{2} \hat{F}_{2h0 2l}. \tag{29a}$$

Group P2$_1$

$$\sum a_j b_j = \sum_{j=1}^{N/2} n_j e^{2\pi i k y_j} \left(e^{2\pi i (h x_j + l z_j)} + (-1)^k e^{-2\pi i (h x_j + l z_j)}\right).$$

We assume

$$a_j = (n_j)^{\frac{1}{2}} e^{2\pi i k y_j}, \quad b_j = (n_j)^{\frac{1}{2}} \left(e^{2\pi i (h x_j + l z_j)} + (-1)^k e^{-2\pi i (h x_j + l z_j)}\right).$$

The inequality acquires the form

$$\hat{F}_{hkl}^2 \leqslant \sum_{j=1}^{N/2} n_j \left| e^{2\pi i k y_j} \right|^2 \sum_{j=1}^{N/2} n_j \left| e^{2\pi i (h x_j + l z_j)} + (-1)^k e^{-2\pi i (h x_j + l z_j)} \right|^2.$$

As before, we discard the first sum, and this strengthens the inequality. The second sum gives

$$2+(-1)^k\left[e^{2\pi i2(hx+lz)}+e^{-2\pi i2(hx+lz)}\right]=2+(-1)^k 2\cos 2\pi(2hx+2lz).$$

Consequently,

$$\hat{F}^2_{hkl}\leqslant\frac{1}{2}2\sum_{j=1}^{N/2}n_j\left[1+(-1)^k\cos 2\pi\left(2hx_j+2lz_j\right)\right],$$

or finally

$$\hat{F}^2_{hkl}\leqslant\frac{1}{2}+\frac{1}{2}(-1)^k\hat{F}_{2h02l}. \qquad (29b)$$

<u>Group P4</u>

$$\sum a_jb_j=2\sum_{j=1}^{N/4}n_j\left[\cos 2\pi\left(hx_j+ky_j\right)+\cos 2\pi\left(-kx_j+hy_j\right)\right]e^{2\pi ilz_j}.$$

Let us assume

$$a_j=(2n_j)^{\frac{1}{2}}e^{2\pi ilz_j};\quad b_j=(2n_j)^{\frac{1}{2}}\left[\cos 2\pi\left(hx_j+ky_j\right)+\right.$$
$$\left.+\cos 2\pi\left(-kx_j+hy_j\right)\right].$$

By substituting into the inequality, discarding $|e^{2\pi ilzj}|^2$ and applying the trigonometric formula

$$(\cos\alpha+\cos\beta)^2=1+\frac{1}{2}\cos 2\alpha+\frac{1}{2}\cos 2\beta+\cos(\alpha+\beta)+\cos(\alpha-\beta),$$

we obtain

$$\hat{F}^2_{hkl}\leqslant\sum_{j=1}^{N/4}n_j\left\{1+\frac{1}{2}\cos 2\pi\left(2hx_j+2ky_j\right)+\right.$$
$$+\frac{1}{2}\cos 2\pi\left(-2kx_j+2hy_j\right)+\cos 2\pi\left[(h-k)x_j+(k+h)y_j\right]+$$
$$\left.+\cos 2\pi\left[(k+h)x_j+(h-k)y_j\right]\right\}.$$

Whence

$$\hat{F}^2_{lkl} \leqslant \frac{1}{4} + \frac{1}{4}\hat{F}_{2h,2k,0} + \frac{1}{2}\hat{F}_{h-k,h+k,0}. \quad (30)$$

<u>Group P3</u>

$$\sum a_j b_j = \sum_{j=1}^{N/3} n_j e^{2\pi i l z_j} \left\{ e^{i2\pi(hx_j+ky_j)} + e^{i2\pi(kx_j+iy_j)} + e^{i2\pi(ix_j+hy_j)} \right\}.$$

We assume

$$a_j = n_j^{\frac{1}{2}} e^{2\pi i l z_j}; \quad b_j = n_j^{\frac{1}{2}} \left\{ e^{i2\pi(hx_j+ky_j)} + e^{i2\pi(kx_j+iy_j)} + e^{i2\pi(ix_j+hy_j)} \right\}.$$

By substituting into the inequality, discarding $|e^{2\pi i l z_j}|$ and performing the operation of multiplying b_j by its complex conjugate, we obtain

$$\hat{F}_{lkil} \leqslant \frac{1}{3} \sum_{j=1}^{N/3} n_j \{3 + 2\cos 2\pi[(h-k)x_j + (k-i)y_j] +$$
$$+ 2\cos 2\pi[(i-h)x_j + (h-k)y_j] + 2\cos 2\pi[(k-i)x_j +$$
$$+ (i-h)y_j]\}$$

or

$$\hat{F}_{hkil} \leqslant \frac{1}{3} + \frac{2}{3} Re\{\hat{F}_{h-k,\, k-i,\, i-h,\, 0}\}. \quad (31)$$

We have examined all the typical cases of groups with one element of symmetry. By proceeding in the same way, the reader can easily derive the remaining formulas in Table 5.

We shall deal separately with the practical applications of inequalities, but shall point out at this time that the method just analyzed leads to inequalities of different usefulness. For the groups P4, $P4_1$ and $P4_2$ inequalities arise with a boundary $\epsilon = \frac{1}{4}$, and for P6, 6_1, 6_2 and 6_3 with $\epsilon = \frac{1}{6}$ (here the ϵ's are the roots of the equations and $x^2 = \frac{1}{4} - \frac{3}{4}x$ and $x^2 = \frac{1}{6} - \frac{5}{6}x$).

These are very strong inequalities (compare page 165). On the other hand, for the groups P3, $P3_1$, $P\bar{4}$, the inequalities which arise are uninteresting in practice [we recall that experiment gives the modulus of a structure amplitude and not its real part R(F)].

TABLE 5. Inequalities for Some Symmetry Groups

Group	Inequality
$P1$	$\hat{F}^2_{hkl} \leqslant \dfrac{1}{2} + \dfrac{1}{2}\,\hat{F}_{2h,\,2k,\,2l}$
$P2$	$\hat{F}^2_{hkl} \leqslant \dfrac{1}{2} + \dfrac{1}{2}\,\hat{F}_{2h02l}$
Pm	$\hat{F}^2_{hkl} \leqslant \dfrac{1}{2} + \dfrac{1}{2}\,\hat{F}_{02k0}$
$P2_1$	$\hat{F}^2_{hkl} \leqslant \dfrac{1}{2} + \dfrac{1}{2}\,(-1)^k\,\hat{F}_{2h02l}$
$P3$	$\hat{F}^2_{hkl} \leqslant \dfrac{1}{3} + \dfrac{2}{3}\,Re\left\{\hat{F}_{h-k,\,2k+h,0}\right\}$
$P3_1$	$\hat{F}^2_{hkl} \leqslant \dfrac{1}{3} + \dfrac{2}{3}\,|\hat{F}_{h-k,\,2k+h,\,0}|\cos\left(\alpha_{h-k,\,2k+h,\,0} + \dfrac{2\pi}{3}l\right)$
$P4$	$\hat{F}^2_{hkl} \leqslant \dfrac{1}{4} + \dfrac{1}{4}\,\hat{F}_{2h,\,2k,\,0} + \dfrac{1}{2}\,\hat{F}_{h-k,h+k,0}$
$P\bar{4}$	$\hat{F}^2_{hkl} \leqslant \dfrac{1}{4} + \dfrac{1}{4}\,\hat{F}_{2h,\,2k,\,0} + \dfrac{1}{2}\,Re\left\{\hat{F}_{h-k,\,h+k,\,2l}\right\}$
$P4_1$	$\hat{F}^2_{hkl} \leqslant \dfrac{1}{4} + \dfrac{1}{4}\,(-1)^l\,\hat{F}_{2h,\,2k,\,0} + \dfrac{1}{2}\,\cos\dfrac{\pi l}{2}\,\hat{F}_{h-k,\,h+k,\,0}$
$P4_2$	$\hat{F}^2_{hkl} \leqslant \dfrac{1}{4} + \dfrac{1}{4}\,\hat{F}_{2h,\,2k,\,0} + \dfrac{1}{2}\,(-1)^l\,\hat{F}_{h-k,\,h+k,\,0}$
$P6$	$\hat{F}^2_{hkl} \leqslant \dfrac{1}{6} + \dfrac{1}{6}\,\hat{F}_{2h,\,2k,\,0} + \dfrac{1}{3}\,\hat{F}_{h-k,\,h+2k,\,0} + \dfrac{1}{3}\,\hat{F}_{hh0}$
$P6_1$	$\hat{F}^2_{hkl} \leqslant \dfrac{1}{6} + \dfrac{1}{6}\,(-1)^l\,\hat{F}_{2h,\,2k,\,0} + \dfrac{1}{3}\,\cos\dfrac{2\pi l}{3}\,\hat{F}_{h-k,\,h+2k,\,0} + {}$ $+ \dfrac{1}{3}\,\cos\dfrac{2\pi l}{6}\,\hat{F}_{hh0}$
$P6_3$	$\hat{F}^2_{hkl} \leqslant \dfrac{1}{6} + \dfrac{1}{6}\,(-1)^l\,\hat{F}_{2h,\,2k,\,0} + \dfrac{1}{3}\,\hat{F}_{h-k,\,h+2k,\,0} + \dfrac{1}{3}\,(-1)^l\,\hat{F}_{hh0}$
$P6_2$	$\hat{F}^2_{hkl} \leqslant \dfrac{1}{6} + \dfrac{1}{6}\,\hat{F}_{2h,\,2k,\,0} + \dfrac{1}{3}\,\cos\dfrac{2\pi l}{3}\,\hat{F}_{h-k,\,h+2k,\,0} + {}$ $+ \dfrac{1}{3}\,\cos\dfrac{2\pi l}{3}\,\hat{F}_{hk0}$

Now let us study space groups which contain several elements of symmetry. The difference, as compared with the preceding examples, lies in the possibility of breaking $a_j b_j$ into factors in several ways that lead to useful inequalities.

We shall present several examples to elucidate the method, thus giving the reader the opportunity to derive the inequalities for the space group that interests him.

Group $P2_1/c$ (C_{2h}^5)

Let us assume $\sum a_j b_j$ to be equal to the unitary structure amplitude

$$\sum a_j b_j = 4\sum_{j=1}^{N/4} n_j \cos 2\pi\left(hx_j + lz_j + \frac{k+l}{4}\right)\cos 2\pi\left(ky_j - \frac{k+l}{4}\right).$$

Two methods of breaking up $a_j b_j$ into factors are possible here

$$a_j = 2n_j^{\frac{1}{2}}, \quad b_j = 2n_j^{\frac{1}{2}} \cos 2\pi\left(hx_j + lz_j + \right.$$

$$\left. + \frac{k+l}{4}\right)\cos 2\pi\left(ky_j - \frac{k+l}{4}\right)$$

$$a_j = 2n_j^{\frac{1}{2}} \cos 2\pi\left(ky_j - \frac{k+l}{4}\right),$$

$$b_j = 2n_j^{\frac{1}{2}} \cos 2\pi\left(hx_j + lz_j + \frac{k+l}{4}\right).$$

Let us make the first substitution, using the trigonometric formula $4\cos^2\alpha\cos^2\beta = 1 + \cos 2\alpha + \cos 2\beta + \cos 2\alpha\cos 2\beta$, and substitute it into the inequality. We obtain

$$\hat{F}_{hkl}^2 \leqslant 4\sum_{j=1}^{N/4} n_j 4\sum_{j=1}^{N/4} n_j \frac{1}{4}\left[1 + \cos 2\pi\left(2hx_j + 2lz_j + \frac{k+l}{2}\right) + \right.$$

$$+ \cos 2\pi\left(2ky_j - \frac{k+l}{2}\right) + \cos\left(2hx_j + 2lz_j + \right.$$

$$\left. + \frac{k+l}{2}\right)\cos\left(2ky_j - \frac{k+l}{2}\right)\Big].$$

Thus,

$$\hat{F}^2_{hkl} \leqslant \frac{1}{4}\left(1 + (-1)^{k+l}\hat{F}_{2h02l} + (-1)^{k+l}\hat{F}_{02k0} + \hat{F}_{2h2k2l}\right). \quad (32)$$

Making the second substitution and using the trigonometric formula $2\cos^2\alpha = 1 + \cos 2\alpha$, we obtain

$$\hat{F}^2_{hkl} \leqslant \frac{1}{4}(1 + (-1)^{k+l}\hat{F}_{2h02l} + (-1)^{k+l}\hat{F}_{02k0} + \hat{F}_{02k0}\hat{F}_{2h02l}). \quad (33)$$

Both formulas are very similar, but the second one is of greater interest, as three and not four different structure amplitudes enter the inequality. It is true that, correspondingly, the boundary ϵ of the first inequality is lower (1/4 as opposed to 1/3 in the second).

If account is taken of the fact that centering of the cell does not bring in anything new, then the inequalities for all the groups of class C_{2h} will be obvious without an independent derivation (which, by the way, presents no difficulty, being fully analogous to the one just given for C^5_{2h}).

For all groups of this class we obtain

$$\hat{F}^2_{hkl} \leqslant \frac{1}{4}[1 + (-1)^{\alpha}\hat{F}_{2h02l} + (-1)^{\alpha}\hat{F}_{02k0} + \hat{F}_{2h02l}\hat{F}_{02k0}]; \quad (34)$$

$$\hat{F}^2_{hkl} \leqslant \frac{1}{4}[1 + (-1)^{\alpha}\hat{F}_{2h02l} + (-1)^{\alpha}\hat{F}_{02k0} + \hat{F}_{2h2k2l}], \quad (35)$$

where α has the following values

$$P2/m \text{ and } C2/m \quad \alpha = 0; \quad P2/c \text{ and } C2/c \quad \alpha = l;$$
$$P2_1/m \quad \alpha = k; \quad P2_1/c \quad \alpha = k + l.$$

Group Pbca

The computations are analogous to the preceding:

$$\sum a_j b_j = 8\sum_{j=1}^{N/8} nj \cos 2\pi\left(hx_j - \frac{h-k}{4}\right) \times$$
$$\times \cos 2\pi\left(ky_j - \frac{k-l}{4}\right)\cos 2\pi\left(lz_j - \frac{l-h}{4}\right).$$

Now there are four useful ways of breaking up $a_j b_j$

$$a_j = (8n_j)^{\frac{1}{2}};$$

$$a_j = (8n_j)^{\frac{1}{2}} \cos 2\pi \left(hx_j - \frac{h-k}{4}\right);$$

$$a_j = (8n_j)^{\frac{1}{2}} \cos 2\pi \left(ky_j - \frac{k-l}{4}\right);$$

$$a_j = (8n_j)^{\frac{1}{2}} \cos 2\pi \left(lz_j - \frac{l-h}{4}\right).$$

The first variation, with the aid of the trigonometric formula

$$\cos^2 \alpha \cos^2 \beta \cos^2 \gamma = \frac{1}{8}(1 + \cos 2\alpha + \cos 2\beta + \cos 2\gamma +$$

$$+ \cos 2\alpha \cos 2\gamma + \cos 2\alpha \cos 2\beta + \cos 2\beta \cos 2\gamma + \cos 2\alpha \cos 2\beta \cos 2\gamma)$$

gives the inequality

$$\hat{F}^2_{hkl} \leqslant 8 \sum_{j=1}^{N/8} n_j \sum_{j=1}^{N/8} n_j \left(1 + \cos 2\pi \left(2hx_j - \frac{h-k}{2}\right) + \right.$$

$$+ \cos 2\pi \left(2ky_j - \frac{k-l}{2}\right) + \cos 2\pi \left(2lz_j - \frac{l-h}{2}\right) +$$

$$+ \cos 2\pi \left(2hx_j - \frac{h-k}{2}\right) \cos 2\pi \left(2ky_j - \frac{k-l}{2}\right) +$$

$$+ \cos 2\pi \left(2hx_j - \frac{h-k}{2}\right) \cos 2\pi \left(2lz_j - \frac{l-h}{2}\right) +$$

$$+ \cos 2\pi \left(2ky_j - \frac{k-l}{2}\right) \cos 2\pi \left(2lz_j - \frac{l-h}{2}\right) +$$

$$+ \cos 2\pi \left(2hx_j - \frac{h-k}{2}\right) \cos 2\pi \left(2ky_j - \frac{k-l}{2}\right) \times$$

$$\left. \times \cos 2\pi \left(2lz_j - \frac{l-h}{2}\right)\right).$$

We note that

$$F_{2h00} = 8 \sum_{j=1}^{N/8} n_j \cos 2\pi (2hx_j),$$

$$\hat{F}_{2h02l} = 8 \sum_{j=1}^{N/8} n_j \cos 2\pi 2hx_j \cos 2\pi 2lz_j$$

etc. Comparing with the sums just obtained, we see that they differ from structure amplitudes only by factors $(-1)^{\alpha}$. Determining α, we obtain

$$\hat{F}^2_{hkl} \leqslant \frac{1}{8}(1 + (-1)^{h-k}\hat{F}_{2h00} + (-1)^{k-l}\hat{F}_{02k0} + (-1)^{l-h}\hat{F}_{002l} + $$
$$+ (-1)^{k-l}\hat{F}_{2h02l} + (-1)^{l-h}\hat{F}_{2h2k0} + (-1)^{h-k}\hat{F}_{02k2l} + \hat{F}_{2h2k2l}). \quad (36)$$

The second variation in breaking $a_j b_j$ into factors with the aid of the formulas $\cos^2\alpha = \frac{1}{2}(1 + \cos 2\alpha)$ and $\cos^2\beta \cos^2\gamma = \frac{1}{4}(1 + \cos 2\beta + \cos 2\gamma + + \cos 2\gamma \cos 2\beta)$ gives

$$\hat{F}^2_{hkl} \leqslant \frac{1}{8}[1 + (-1)^{h-k}\hat{F}_{2h00} + (-1)^{k-l}\hat{F}_{02k0} + (-1)^{l-h}\hat{F}_{002l} +$$
$$+ (-1)^{h-k}\hat{F}_{02k2l} + (-1)^{h-l}\hat{F}_{2h00}\hat{F}_{02k0} +$$
$$+ (-1)^{l-k}\hat{F}_{2h00}\hat{F}_{002l} + \hat{F}_{2h00}\hat{F}_{02k2l}]. \quad (37)$$

Finally, the last two variations are these

$$\hat{F}^2_{hkl} \leqslant \frac{1}{8}[1 + (-1)^{h-k}\hat{F}_{2h00} + (-1)^{k-l}\hat{F}_{02k0} + (-1)^{l-h}F_{002l} +$$
$$+ (-1)^{k-l}\hat{F}_{2h02l} + (-1)^{h-l}\hat{F}_{2h00}\hat{F}_{02k0} +$$
$$+ (-1)^{k-h}\hat{F}_{02k0}\hat{F}_{002l} + \hat{F}_{02k0}\hat{F}_{2h02l}]; \quad (38)$$

$$\hat{F}^2_{hkl} \leqslant \frac{1}{8}[1 + (-1)^{h-k}\hat{F}_{2h00} + (-1)^{k-l}\hat{F}_{02k0} +$$
$$+ (-1)^{l-h}\hat{F}_{002l} + (-1)^{l-h}\hat{F}_{2h2k0} + (-1)^{k-h}\hat{F}_{002l}\hat{F}_{02k0} +$$
$$+ (-1)^{l-k}F_{2h00}\hat{F}_{002l} + \hat{F}_{002l}\hat{F}_{2h2k0}]. \quad (39)$$

The formulas for the other space groups of class D_{2h} are almost identical [the difference is in the exponents of (-1)].

The formulas just derived have the boundary $\epsilon = 1/8$. Thus, success, even partial (prohibition of some sign correlations) in their application is guaranteed as a rule. The application of these inequalities is especially expedient if some of the signs are already known.

Let us notice that the application of an inequality leads to the determination of positive signs of the addends. Since the sign of an amplitude with an even index coincides with the sign of a degenerate structure product, we see here also that inequalities may either give no result, or else indicate that this structure product is greater than zero (see page 157).

It is essential to emphasize that the formulas derived for these or other space groups, are also correct for other more symmetrical groups if they contain as subgroups the groups under study. Let us say that, for all centrosymmetric crystals, the formulas derived for $P\bar{1}$ are correct, and for group Pbca those derived for $P2_1/a$, etc.

8. The Second Method of Applying Cauchy's Inequality

The second method consists in equating the sum $\sum a_j b_j$ — the left side of Cauchy's inequality — to the sum and the difference of two structure amplitudes:

$$\sum a_j b_j = \hat{F}_{hkl} \pm \hat{F}_{h'k'l'}. \qquad (40)$$

Let us see what the use of this method leads to for a few space groups:

Group $P\bar{1}$

$$\sum a_j b_j = \sum_{j=1}^{N/2} n_j (\cos \alpha_j \pm \cos \alpha_j'),$$

i.e., for the + sign

$$\sum a_j b_j = 2 \sum_{j=1}^{N/2} 2n_j \cos \frac{(\alpha_j + \alpha_j')}{2} \cos \frac{(\alpha_j - \alpha_j')}{2}$$

and for the − sign

$$\sum a_j b_j = -2 \sum_{j=1}^{N/2} 2n_j \sin \frac{(\alpha_j + \alpha_j')}{2} \sin \frac{(\alpha_j - \alpha_j')}{2}.$$

Assuming a_j to be equal to $2n_j^{\frac{1}{2}} \cos \frac{\alpha_j - \alpha_j'}{2}$, or in the second case to $2n_j^{\frac{1}{2}} \sin \frac{\alpha_j + \alpha_j}{2}$, substituting into Cauchy's inequality, and using the formulas $2\cos^2\alpha = 1 + \cos 2\alpha$ and $2\sin^2\alpha = 1 - \cos 2\alpha$, we obtain

$$(\hat{F}_{\mathbf{H}} \pm \hat{F}_{\mathbf{H}'})^2 \leqslant (1 \pm \hat{F}_{\mathbf{H}+\mathbf{H}'})(1 \pm \hat{F}_{\mathbf{H}-\mathbf{H}'}). \tag{41}$$

These two inequalities, correct for all centrosymmetric crystals, have the boundary $\epsilon = \frac{1}{3}$, and are very convenient for practical use.

Group $P2_1/a$

$$\sum a_j b_j = 4\sum_{j=1}^{N/4} n_j \left[\cos 2\pi\left(hx_j + lz_j + \frac{h+k}{4}\right)\cos 2\pi\left(ky_j - \frac{h+k}{4}\right) \pm \right.$$
$$\left. \pm \cos 2\pi\left(h'x_j + l'z_j + \frac{h'+k'}{4}\right)\cos 2\pi\left(k'y_j - \frac{h'+k'}{4}\right)\right].$$

Let us study only the simplest division into factors, i.e., assume $a_j = 2n_j^{\frac{1}{2}}$. Then Cauchy's inequality takes the form

$$(\hat{F}_{hkl} \pm \hat{F}_{h'k'l'})^2 \leqslant 4\sum_{j=1}^{N/4} n_j \left[\cos 2\pi\left(hx_j + lz_j + \frac{h+k}{4}\right) \times \right.$$
$$\times \cos 2\pi\left(ky_j - \frac{h+k}{4}\right) \pm \cos 2\pi\left(h'x_j + l'z_j + \frac{h'+k'}{4}\right) \times$$
$$\left. \times \cos 2\pi\left(k'y_j - \frac{h'+k'}{4}\right)\right]^2.$$

Using the following trigonometric formulas:

$$\cos^2\alpha \cos^2\beta = \frac{1}{4}(1 + \cos 2\alpha + \cos 2\beta + \cos 2\alpha \cos 2\beta),$$

$$\cos^2\alpha' \cos^2\beta' = \frac{1}{4}(1 + \cos 2\alpha' + \cos 2\beta' + \cos 2\alpha' \cos 2\beta'),$$

$$4\cos\alpha\cos\alpha'\cos\beta\cos\beta' = \cos(\alpha+\alpha')\cos(\beta+\beta') +$$
$$+ \cos(\alpha-\alpha')\cos(\beta+\beta') + \cos(\alpha+\alpha')\cos(\beta-\beta') +$$
$$+ \cos(\alpha-\alpha')\cos(\beta-\beta'),$$

we obtain

$$
\begin{aligned}
(\hat{F}_{hkl} \pm \hat{F}_{h'k'l'})^2 \leqslant \frac{1}{4} \{ 2 + (-1)^{h+k} \hat{F}_{2h02l} + (-1)^{h+k} \hat{F}_{02k0} + \hat{F}_{2h2k2l} + \\
+ (-1)^{h'+k'} \hat{F}_{2h'02l'} + (-1)^{h'+k'} \hat{F}_{02k'0} + \hat{F}_{2h'2k'2l'} \pm 2 [\hat{F}_{h+h', k+k', l+l'} + \\
+ (\hat{F}_{h-h', k+k', l-l'} + \hat{F}_{h+h', k-k', l+l'}) (-1)^{h'+k'} + \hat{F}_{h-h', k-k', l-l'}] \}. \qquad (42)
\end{aligned}
$$

The boundary of application of this inequality lies at $\epsilon = {}^1/_8$. Thus, such an inequality fails to give information on the correlation of signs only in the case that all the structure amplitudes that enter into it are less than 0.125. If all the amplitudes are within the limits 0.1 – 0.2, then the inequality is sure to give some kind of sign information.

Of course, the practical analysis of an inequality containing twelve different structure amplitudes is very cumbersome. It is practical to use the inequality only in case the greater part of the signs of the amplitudes which enter into it are already known. The construction of such an inequality for this or some other space group is not a complicated problem.[3] The analysis of an inequality containing a large number of variables is much more difficult. If not one of the signs is known, then it is hardly possible to use in practice an inequality that contains more than four or five variables.

The derivation of correlations between the quantities $\hat{F}$ is of course possible by using other algebraic inequalities too. However, the inequalities analyzed in these two paragraphs result from the general equation of connection (48), and therefore are "the best" inequalities. For example, all the inequalities derived by Gillis [39] have inferior possibilities in comparison with the inequalities that follow from (48).

9. The Fundamental Equation Connecting Structure Amplitudes

We shall arrive at the most general connection between structure amplitudes in the following manner [40].

Let us consider $\underline{m}$ arbitrary nodes of the reciprocal lattice of a crystal. Let the corresponding reciprocal vectors be

$$\mathbf{H}_p \, (p = 1, 2, \ldots, m).$$

[3]One should note the great convenience of computations carried out by substituting trigonometric functions for complex ones. By this method the trigonometric formulas necessary for the derivation of the inequality are easily derived.

To each node of the reciprocal lattice we shall assign an N-dimensional vector

$$G_p = \sum_{j=1}^{N} e_j n_j^{\frac{1}{2}} \exp 2\pi i H_p r_j, \qquad (43)$$

where the e_j are the orthonormal unit vectors of the N-dimensional space, and r_j and n_j, as always, are the radius vectors and "weights" of the atoms in the cell.

Since the n_j are positive numbers, the $n_j^{\frac{1}{2}}$ are real numbers. From this some of the properties of the vectors **G** follow. First of all the vectors **G** are unitary vectors, inasmuch as

$$(G_p, G_p) = \sum_{j=1}^{N} n_j = 1, \qquad (44)$$

and furthermore, the scalar product

$$(G_p, G_q) = \sum_{j=1}^{N} n_j \exp (H_p - H_q) r_j = \hat{F}_{pq}, \qquad (45)$$

obtained by multiplying by the complex conjugate, is but the structure amplitude of the node $H_p - H_q$ (denoted hereinafter in abbreviated form as $\hat{F}_{pq}$).

As is known ([4], vol. III, ch. 1, p. 61) the determinant of order m composed of the scalar products

$$D_m = \begin{vmatrix} (G_1, G_1) & (G_1, G_2) & \vdots & (G_1, G_m) \\ (G_2, G_1) & (G_2, G_2) & \vdots & (G_2, G_m) \\ (G_3, G_1) & (G_3, G_2) & \vdots & (G_3, G_m) \\ \cdots & \cdots & \cdots & \cdots \\ (G_m, G_1) & (G_m, G_2) & \vdots & (G_m, G_m) \end{vmatrix} \qquad (46)$$

is called the Gram determinant of the vectors G_p (p = 1...m). This determinant has the following wonderful properties, proved, for example, in the reference cited.

1. The Gram determinant of any system of m vectors cannot be negative.

$$D_m \geqslant 0. \qquad (47)$$

2. If the vectors $\mathbf{G}_p$ are linearly dependent, as in the case when the number of dimensions of the space is smaller then $\underline{m}$, then the Gram determinant is equal to zero.

Forming the Gram determinant (46) from the vectors (43) we obtain the fundamental equation connecting structure amplitudes:

$$D_m = \begin{vmatrix} 1 & \hat{F}_{12} & \hat{F}_{1m} \\ \hat{F}_{21} & 1 & \hat{F}_{2m} \\ \hat{F}_{31} & \hat{F}_{32} & \hat{F}_{3m} \\ \cdots & \cdots & \cdots \\ \hat{F}_{m1} & \hat{F}_{m2} & 1 \end{vmatrix} \geqslant 0. \tag{48}$$

Moreover, this inequality is correct for $m \leq N$. Schwartz' inequality is a particular case of (47).

Equation (48) gives the most general connection between the amplitudes and deserves the name we propose to give it.

If a crystal is centrosymmetric, $\hat{F}_{pq} = \hat{F}_{qp}$. Then $\frac{m(m-1)}{2}$ different amplitudes enter into the determinant of the $\underline{m}$th order. Forming D_m for these structure amplitudes, we may arbitrarily choose $(m-1)$ of them, let us say $\hat{F}_{1q}$. The indices of the remaining amplitudes $\hat{F}_{pq}$ are determined by the indices of these first $(m-1)$ amplitudes.

Let us construct, for example, D_4 for a centrosymmetric crystal. For the initial ones, let us take the nodes of the reciprocal lattice with the indices 125, $31\bar{1}$ and 040. Then

$$D_4 = \begin{vmatrix} 1 & & & \\ \hat{F}_{125} & 1 & & \\ \hat{F}_{31\bar{1}} & \hat{F}_{2\bar{1}\bar{6}} & 1 & \\ F_{040} & \hat{F}_{\bar{1}2\bar{5}} & \hat{F}_{\bar{3}3\bar{1}} & 1 \end{vmatrix}. \tag{49}$$

Let us take note of the geometric meaning which structure amplitudes have acquired in our concept: The unitary $\hat{F}$'s are the cosines of the angles formed between the N-dimensional vectors (43).

The equation of connection (48) includes the Harker and Kasper inequalities, Banerjee's equations, and all the other proposed correlations between amplitudes.

The problem of determining the signs of the amplitude for a centrosymmetric crystal is soluble in principle. The limitations on the practical application of (48) will be elucidated in the paragraphs that follow.

Equation (48) was first derived by Goedkoop [42] on the basis of the previous articles by Karle and Hauptman. The development of (48) in these articles was different; the authors did not indicate any geometrical meaning for the basic equation connecting structure amplitudes.

10. Form of the Connecting Determinants of Low Orders

To show the generality of (48) and the methods of obtaining different inequalities or equalities with the aid of the basic equation of connection, let us first of all analyze in detail the form of such determinants of low orders. We shall limit ourselves to centrosymmetric crystals.

m = 2.

$$D_2 = \begin{vmatrix} 1 & \hat{F}_H \\ \hat{F}_H & 1 \end{vmatrix}. \tag{50}$$

Equation (48) gives the trivial result $F_H^2 \leq 1$.

m = 3.

In the general case

$$D_3 = \begin{vmatrix} 1 & \hat{F}_H & \hat{F}_K \\ \hat{F}_H & 1 & \hat{F}_{K-H} \\ \hat{F}_K & \hat{F}_{K-H} & 1 \end{vmatrix}. \tag{51}$$

From (48) it follows that

$$1 - \hat{F}_H^2 - \hat{F}_K^2 - \hat{F}_{K-H}^2 + 2\hat{F}_H \hat{F}_K \hat{F}_{K-H} \geqslant 0. \tag{52}$$

Taking the amplitude $\hat{F}_{\bar{H}}$ (instead of $\hat{F}_H$) as the initial one, and taking into account that for a centrosymmetric crystal, $\hat{F}_H = \hat{F}_{\bar{H}}$ we also obtain

$$1 - \hat{F}_H^2 - \hat{F}_K^2 - \hat{F}_{K+H}^2 + 2\hat{F}_H \hat{F}_K \hat{F}_{K+H} \geqslant 0.$$

If the number of atoms in the cell is equal to N = 2, then (52) becomes an equality.

The boundary of (52) is $\epsilon = 1/2$.

Even in this simple case we can obtain additional equations from the determinant of the third order connecting **H** and **K.** Assuming, for example, K = $\bar{\mathbf{H}}$, we obtain (28):

$$\hat{F}^2_{\mathrm{H}} \leqslant \frac{1}{2} + \frac{1}{2}\hat{F}_{2\mathrm{H}}.$$

In fact, substitution **K** = $\bar{\mathbf{H}}$ in (52) we obtain

$$1 - \hat{F}^2_{2\mathrm{H}} - 2\hat{F}^2_{\mathrm{H}}(1 - \hat{F}_{2\mathrm{H}}) \geqslant 0,$$

but since $(1 - \hat{F}_{2\mathrm{H}})$ is a positive number, the left side of the inequality may be divided by it. Thus, we come to (28).

Let us apply the same determinant to a space group containing the axis 2. Assuming that **H** = $h\bar{k}l$ and **K** = hkl, we obtain

$$D_3 = \begin{vmatrix} 1 & \hat{F}_{h\bar{k}l} & \hat{F}_{hkl} \\ \hat{F}_{h\bar{k}l} & 1 & \hat{F}_{02k0} \\ \hat{F}_{hkl} & \hat{F}_{02k0} & 1 \end{vmatrix}.$$

Repeating the same computation we arrive at (29a). Therefore, the first four inequalities of the table on page 144 are particular cases of the inequality $D_3 \geq 0$.

m = 4.

In the general case

$$D_4 = \begin{vmatrix} 1 & \hat{F}_{\mathrm{H}} & \hat{F}_{\mathrm{K}} & \hat{F}_{\mathrm{L}} \\ \hat{F}_{\mathrm{H}} & 1 & \hat{F}_{\mathrm{K-H}} & \hat{F}_{\mathrm{L-H}} \\ \hat{F}_{\mathrm{K}} & \hat{F}_{\mathrm{K-H}} & 1 & \hat{F}_{\mathrm{L-K}} \\ \hat{F}_{\mathrm{L}} & \hat{F}_{\mathrm{L-H}} & \hat{F}_{\mathrm{L-K}} & 1 \end{vmatrix},$$

i.e., six different structure amplitudes enter into the connecting determinant.

$$D_4 = 1 - \hat{F}^2_{\mathrm{H}} - \hat{F}^2_{\mathrm{K}} - \hat{F}^2_{\mathrm{L}} - \hat{F}^2_{\mathrm{L-K}} - \hat{F}^2_{\mathrm{K-H}} - \hat{F}^2_{\mathrm{L-H}} + \hat{F}^2_{\mathrm{H}}\hat{F}^2_{\mathrm{L-K}} + \hat{F}^2_{\mathrm{K}}\hat{F}^2_{\mathrm{L-H}} + \hat{F}^2_{\mathrm{L}}\hat{F}^2_{\mathrm{K-H}} + 2\hat{F}_{\mathrm{H}}\hat{F}_{\mathrm{K}}\hat{F}_{\mathrm{K-H}} + 2\hat{F}_{\mathrm{H}}\hat{F}_{\mathrm{L}}\hat{F}_{\mathrm{L-H}} + 2\hat{F}_{\mathrm{K}}\hat{F}_{\mathrm{L}}\hat{F}_{\mathrm{L-K}} + 2\hat{F}_{\mathrm{K-H}}\hat{F}_{\mathrm{L-K}}\hat{F}_{\mathrm{L-H}} - 2\hat{F}_{\mathrm{H}}\hat{F}_{\mathrm{L}}\hat{F}_{\mathrm{L-K}}\hat{F}_{\mathrm{K-H}} - 2\hat{F}_{\mathrm{H}}\hat{F}_{\mathrm{K}}\hat{F}_{\mathrm{L-K}}\hat{F}_{\mathrm{L-H}} - 2\hat{F}_{\mathrm{L}}\hat{F}_{\mathrm{K}}\hat{F}_{\mathrm{K-H}}\hat{F}_{\mathrm{L-H}}. \quad (53)$$

The determinant of the fourth order gives $D_4 = 0$ when $N = 3$. But, since the computation is carried out for a centrosymmetric crystal, the equating to zero of this expression will be correct only when $N = 2$.

Let us take note of the following circumstance (see page 158) common to all connecting determinants: the value of the connecting determinant is defined by the absolute magnitudes of the structure amplitudes and the signs of the structure products. For (53) this is obvious. Let s_{HK}, s_{HL} and s_{KL} be the signs of the third-order structure products $X_{H\bar{K}}$, $X_{H\bar{L}}$ and $X_{K\bar{L}}$.

Then the inequality

$$D_4 \geqslant 0 \tag{54}$$

can be written in the form

$$\begin{gathered}D_4 = 1 - \hat{F}^2_H - \hat{F}^2_K - \hat{F}^2_L - \hat{F}^2_{L-K} - \hat{F}^2_{K-H} - \hat{F}^2_{L-H} + \hat{F}^2_H \hat{F}^2_{L-K} + \\ + \hat{F}^2_K \hat{F}^2_{L-H} + \hat{F}^2_L \hat{F}^2_{K-H} + 2s_{HK} |\hat{F}_H \hat{F}_K \hat{F}_{H-K}| + 2s_{HL} |\hat{F}_H \hat{F}_L \hat{F}_{H-L}| + \\ + 2s_{KL} |\hat{F}_H \hat{F}_L \hat{F}_{L-K}| + 2s_{HK} s_{KL} s_{HL} |\hat{F}_{K-H} \hat{F}_{L-K} \hat{F}_{L-H}| - \\ - 2s_{HK} s_{KL} |\hat{F}_H \hat{F}_L \hat{F}_{L-K} \hat{F}_{K-H}| - 2s_{HL} s_{KL} |\hat{F}_H \hat{F}_K \hat{F}_{L-K} \hat{F}_{L-H}| - \\ - 2s_{HK} s_{HL} |\hat{F}_L \hat{F}_K \hat{F}_{K-H} \hat{F}_{L-H}| \geqslant 0.\end{gathered} \tag{55}$$

Thus, three signs determine D_4 for known absolute values of the structure amplitudes.

There are already practical difficulties in the analysis of (54), since eight variations—eight sign results—correspond to the three signs.

Let us examine some particular cases of imposing connections on **H**, **K**, and **L**.

<u>a) Linear connections.</u> $\mathbf{L} = \mathbf{H} + \mathbf{K}$. Substitution in (55) brings us to the inequality

$$\begin{gathered}(1 - \hat{F}^2_{H+K})(1 - \hat{F}^2_{H-K}) + (\hat{F}^2_H - \hat{F}^2_K)^2 \geqslant \\ \geqslant 2(\hat{F}^2_H + \hat{F}^2_K)(1 + s_1 s_2 |\hat{F}_{H+K} \hat{F}_{H-K}|) - \\ - 4s_1 |\hat{F}_H \hat{F}_K \hat{F}_{H+K}| - 4s_2 |\hat{F}_H \hat{F}_K \hat{F}_{H-K}|,\end{gathered} \tag{56}$$

where s_1 and s_2 are the signs of the structure products $\hat{F}_H\hat{F}_K\hat{F}_{H+K}$ and $\hat{F}_H\hat{F}_K\hat{F}_{H-K}$, respectively.

We have obtained a convenient and simple inequality for practical application. The methods and the results of its application will be studied in detail below.

Other connections of the type $\mathbf{L} = \alpha\mathbf{H} + \beta\mathbf{K}$ will not lead to simplification of (55).

<u>b) Symmetrical connections.</u> It is convenient to simplity (56) for any crystal containing an axis 2 or 2_1, assuming that $\mathbf{H} = hkl$ and $\mathbf{K} = h\bar{k}l$. Then (for example, for the axis 2_1) we obtain

$$\begin{aligned}(1-\hat{F}^2_{2h02l})(1-\hat{F}^2_{02k0}) \geqslant 4\hat{F}^2_{hkl}(1+s_1s_2\,|\hat{F}_{2h02l}\hat{F}_{02k0}|)- \\ -4\hat{F}^2_{hkl}(-1)^{h+k}\{s_1|\hat{F}_{2h02l}|+s_2|\hat{F}_{02k0}|\}.\end{aligned} \qquad (57)$$

Inequality (57) helps to find the signs s_1 and s_2 of $\hat{F}_{2h02l}$ and $\hat{F}_{02k0}$. This is not contrary to the general theorem just expressed, since, in the case of a degenerate structure product, its sign coincides with the sign of a structure amplitude with even indices.

Inequality (57) is easily simplified and reduced to (33) by dividing it by the quantity $4\,(1-\hat{F}_{2h02l})\,(1-\hat{F}_{02k0})$, which is always positive.

This circumstance is common to all connecting equations: Inequalities derived on the basis of Cauchy's inequality follow from the general connecting equation (48).

Cauchy's inequalities preserve all the advantages of the easiest and quickest derivation of the relationships between structure amplitudes for the case of a symmetric connection.

To obtain, for example, the simple inequality (32), without applying Cauchy's inequality, the determinant D_5 must be computed taking hkl, $\bar{h}\bar{k}\bar{l}$, $h\bar{k}l$ and $\bar{h}k\bar{l}$, for the initial amplitudes. To obtain (42), the determinant must be computed with hkl, $\bar{h}k\bar{l}$, $h\bar{k}l$, $hk\bar{l}$, $h'k'l'$, $\bar{h}'k'\bar{l}'$, $h'\bar{k}'l'$, $h'k'\bar{l}'$ as the initial amplitudes.

It is obvious that it is simpler to use Cauchy's inequality. However, it must be remembered that the connecting equations for amplitudes, the indices of which are not related by symmetry, may be derived only with the aid of (48) by computing the determinants of the corresponding orders.

11. The Determination of Signs

We have shown in the preceding paragraph for D_3 and D_4 that the magnitude of the connecting determinant is defined by the signs of the structure products. Now, we shall prove this important condition for any D_m: The magnitude of the connecting determinant is defined not by the signs of the structure amplitudes, but by those of the structure products (see Section 16).

Let us write the connecting determinant as the sum

$$D_m = \sum (-1)^k \hat{F}_{1\alpha} \hat{F}_{2\beta} \ldots \hat{F}_{m\omega}, \tag{58}$$

where α, β,..., ω run through all the possible m! permutations of the numbers 1,2,...,m, and k is the number of inversions in each permutation.

Since the connecting determinant has unities along its diagonal, then addends will enter into (58) in which all $\hat{F}_{ik}$ are unity, (m − 2) of the elements are unity, (m − 3) of the elements are unity, and so forth.

The first addend in D_m is equal to unity ($\alpha = 1$, $\beta = 2, \ldots, \omega = m$). Addends in which m − 1 of the elements are unity are impossible in (60). In fact, if we assume that $\beta = 2, \ldots, \omega = m$, then we have only the index 1 at our disposal, which transforms $\hat{F}_{1\alpha}$ into unity.

Let us examine now the addends in (58), (m − 2) of the elements of which are unity. These will be addends of the type $\hat{F}_{12}\hat{F}_{21}$, $\hat{F}_{23}\hat{F}_{32}$ etc. Since $\hat{F}_{ki} = \hat{F}_{ik}$, the addends will be the squares of $\hat{F}$. The number of inversions k for addends of this type is equal to unity; therefore the second term enters into D_m with a minus sign. The number of terms in the sum is equal to that of the number of ways of choosing two elements from m, i.e., equal to $\frac{m!}{(m-2)!2!}$. Thus, the second term in D_m can be written in the form

$$-\frac{1}{2} \sum_{i \neq k} \sum \hat{F}_{ik}^2 \tag{59}$$

The terms $\hat{F}_{ik}^2$ and $\hat{F}_{ki}^2$ enter into this sum; therefore the sum must be divided by 2.

Let us examine now the addends in which (m − 3) of the elements are unity. These will be of the type $\hat{F}_{12}\hat{F}_{23}\hat{F}_{31}$, $\hat{F}_{12}\hat{F}_{24}\hat{F}_{41}, \ldots, \hat{F}_{57}\hat{F}_{79}\hat{F}_{95}, \ldots$. Let the unity be chosen from all the rows (and columns), except ijk. For this choice, we have addends $\hat{F}_{ij}\hat{F}_{jk}\hat{F}_{ki}$, $\hat{F}_{ik}\hat{F}_{ji}\hat{F}_{kj}$.

These addends are equal to one another both in magnitude and sign. The order of these indices is established by two inversions (for example, $\hat{F}_{57}\hat{F}_{79}\hat{F}_{95}$ after the first inversion has the form $\hat{F}_{55}\hat{F}_{79}\hat{F}_{97}$, and after the second inversion the form $\hat{F}_{55}\hat{F}_{77}\hat{F}_{99}$). Consequently, these addends enter into the sum with a positive sign. The number of ways of choosing three rows from m is equal to $\frac{m!}{(m-3)!3!}$. Thus, the third term of D_m consists of $\frac{m!}{(m-3)!3!}$ pairs of addends of the type just written and has the form

$$+\sum\sum_{i\neq j\neq k}\sum \hat{F}_{ij}\hat{F}_{jk}\hat{F}_{ki}. \tag{60}$$

The addends of this sum are structure products in the sense defined in Chapter I. Actually, $\hat{F}_{ij}=\hat{F}_{\mathbf{H}_i-\mathbf{H}_j}$, $\hat{F}_{jk}=\hat{F}_{\mathbf{H}_j-\mathbf{H}_k}$ and $\hat{F}_{ki}=\hat{F}_{\mathbf{H}_k-\mathbf{H}_i}$; $(\mathbf{H}_i-\mathbf{H}_j)+(\mathbf{H}_j-\mathbf{H}_k)=-(\mathbf{H}_k-\mathbf{H}_i)$. Therefore, $\hat{F}_{ij}\hat{F}_{jk}\hat{F}_{ki}$ is a structure product.

Let us call "basic" those structure products which are composed of elements of the first column, i.e., products with i = 1: $\hat{F}_{1j}\hat{F}_{jk}\hat{F}_{k1}$. Let us call the signs of these structure products the d e f i n i n g ones and denote them by s_{jk}.

It is not hard to see that this definition is quite justifiable, since the sign of any other structure product may be expressed through the defining signs, that is

$$\operatorname{sign}\ (\hat{F}_{ij}\hat{F}_{jk}\hat{F}_{ki})=s_{ij}s_{jk}s_{ki},$$

which is obvious from

$$\operatorname{sign}\ (\hat{F}_{ij}\hat{F}_{jk}\hat{F}_{ki})=\operatorname{sign}\ (\hat{F}^2_{1i}\hat{F}^2_{1j}\hat{F}^2_{1k}\cdot\hat{F}_{ij}\hat{F}_{jk}\hat{F}_{ki})=$$
$$=\operatorname{sign}\ (\hat{F}_{1i}\hat{F}_{ij}\hat{F}_{j1}\cdot\hat{F}_{1j}\hat{F}_{jk}\hat{F}_{k1}\cdot\hat{F}_{1k}\hat{F}_{ki}\hat{F}_{i1})=s_{ij}s_{jk}s_{ki}.$$

The number of pairs of fundamental structure products is equal to the number of possible combinations of (m − 1) things taken two at a time, i.e., $\frac{(m-1)!}{(m-3)!2!}=\frac{(m-1)(m-2)}{2}$. The signs of the remaining structure products are expressed through the defining signs. Thus, for example, if m = 5, then six signs determine four more signs, if m = 10, then 36 signs determine 84 more signs, if m = 20, then 171 signs determine 969 more signs.

In the general case $\frac{(m-1)(m-2)}{2}$ signs determine $\frac{m!}{(m-3)!3!} - \frac{(m-1)(m-2)}{2} = \frac{(m-1)!}{6(m-4)!}$ more signs.

Let us note that, in general, a total of $\frac{m(m-1)}{2}$ different structure amplitudes enter into D_m, i.e., $(m-1)$ more than the number of defining signs.

It follows from this that, if the defining signs of the structure products are found experimentally (by analysis of the connecting determinant), then, knowing in addition $(m-1)$ signs of the amplitudes (say, $\hat{F}_{1q}$), it is possible to find the signs of all the others. Or else, finding experimentally the signs of the structure products allows one in principle to express the signs of all the amplitudes through the $(m-1)$ signs of the amplitudes in the first column.

We have run somewhat ahead with this formulation, as it must still be shown that the signs of the remaining terms of the determinant are expressed through defining signs. This, however, is quite obvious. The reason is that any term in (58) is composed of elements the column and row indices of which are the same. Let us demonstrate the truth of the above by random examples;

$$\operatorname{sign}\ (\hat{F}_{13}\hat{F}_{24}\hat{F}_{32}\hat{F}_{45}\hat{F}_{51}) =$$

$$= \operatorname{sign}\ (\hat{F}_{13}\hat{F}_{\underline{31}} \cdot \hat{F}_{\underline{12}}\hat{F}_{24}\hat{F}_{\underline{41}} \cdot \hat{F}_{\underline{13}}\hat{F}_{32}\hat{F}_{\underline{21}} \cdot \hat{F}_{\underline{14}}\hat{F}_{45}\hat{F}_{\underline{51}} \cdot \hat{F}_{\underline{15}}\hat{F}_{51}) =$$

$$= s_{24}s_{32}s_{45}.$$

The squares occurring in the parentheses are underlined.

$$\operatorname{sign}\ (\hat{F}_{2j}\hat{F}_{32}\hat{F}_{j3}\hat{F}_{ki}\hat{F}_{ik}\hat{F}_{pq}\hat{F}_{qp}) =$$

$$= \operatorname{sign}\ (\hat{F}_{12}\hat{F}_{2j}\hat{F}_{j1}\hat{F}_{13}\hat{F}_{32}\hat{F}_{21}\hat{F}_{1j}\hat{F}_{j3}\hat{F}_{31}) = s_{2j}s_{32}s_{j3}$$

etc.

Thus, a change of sign in only one structure product brings with it a substantial change in the magnitude of the terms in (58), and, consequently, in the value of D_m.

We have seen that the sign of a large number of structure products may be expressed through the signs of the "fundamental" products. If we intend to determine the signs of structure amplitudes by the statistical method (see Chapter III), then we must select "independent" ones from the structure products constructed. A structure product will be called dependent, if its sign is equal to the sign of the product of three structure products:

$$S_{\mathbf{H}_1+\mathbf{H}_2,\,\overline{\mathbf{H}}_1+\overline{\mathbf{H}}_3,\,\mathbf{H}_2-\mathbf{H}_3} = S_{\mathbf{H}_1,\,\mathbf{H}_2,\,\mathbf{H}_1+\mathbf{H}_2}\,S_{\mathbf{H}_1,\,\mathbf{H}_3,\,\mathbf{H}_1+\mathbf{H}_3}\,S_{\mathbf{H}_2,\,\overline{\mathbf{H}}_3,\,\mathbf{H}_2-\mathbf{H}_3}.$$

Thus, writing out the structure products in an array of three columns of indices,

$$\begin{array}{lll} \mathbf{H}_1 & \mathbf{K}_1 & \mathbf{H}_1+\mathbf{K}_1 \\ \mathbf{H}_2 & \mathbf{K}_2 & \mathbf{H}_2+\mathbf{K}_2 \\ \mathbf{H}_3 & \mathbf{K}_3 & \mathbf{H}_3+\mathbf{K}_3 \\ \cdot\;\cdot & \cdot\;\cdot\;\cdot & \cdot\;\cdot\;\cdot\;\cdot \end{array}$$

the structure products formed of three indices already belonging to the same column should be discarded.

12. Capabilities of the Basic Connecting Equations for Determining Signs

We have already indicated that a simple criterion for the applicability of an inequality to finding signs is the "boundary" ϵ, which has the following meaning: If among the amplitudes that enter into the inequality, there is not one that is larger than ϵ, then the inequality is satisfied by any combination of signs.

Now, we shall show that the boundary of an inequality is determined by the order of the determinant from the positivity of which this inequality originates.

To judge the applicability of an inequality we resort to the following method. Let us assume that all the values of the structure amplitudes – the elements of the determinant – have the same magnitude and are equal to $\hat{F}$.

The determinant takes the form

$$\hat{F}^m \begin{vmatrix} (1+x) & 1 & . & 1 \\ 1 & (1+x) & . & 1 \\ \cdot & \cdot & \cdot & \cdot \\ 1 & 1 & . & (1+x) \end{vmatrix},$$

where $\underline{m}$ is the order of the determinant, and $x = \left(\frac{1}{\hat{F}} - 1\right)$.

Such a determinant is a polynomial of the $\underline{m}$th degree in $\underline{x}$ and is expressed by the formula (see [41], ch. 1, p. 28).

$$\begin{vmatrix} (1+x) & 1 & 1 \ldots & 1 \\ 1 & (1+x) & 1 \ldots & 1 \\ \cdot & \cdot & \cdot & \cdot \\ 1 & 1 & 1 \ldots & (1+x) \end{vmatrix} = x^m + S_1 x^{m-1} + S_2 x^{m-2} + \ldots + S_m,$$

and the coefficients S_k are given by the expression

$$S_k = \frac{1}{k!} \sum_{q_1=1}^{m} \sum_{q_2=1}^{m} \cdots \sum_{q_k=1}^{m} \begin{vmatrix} \hat{F}_{q_1 q_1} \hat{F}_{q_1 q_2} \ldots \hat{F}_{q_1 q_k} \\ \hat{F}_{q_2 q_1} \hat{F}_{q_2 q_2} \ldots \hat{F}_{q_2 q_k} \\ \cdot \quad \cdot \quad \cdot \quad \cdot \quad \cdot \\ \hat{F}_{q_k q_1} \hat{F}_{q_k q_2} \ldots \hat{F}_{q_k q_k} \end{vmatrix}.$$

However, for our determinant all the $\hat{F}_{q_1 q_k} = 1$ and, consequently, all the $S_k = 0$, except $S_1 = m$. The determinant is represented by the formula $x^m + mx^{m-1}$ and, consequently,

$$\begin{vmatrix} 1 & \hat{F}_{12} & \hat{F}_{13} & \cdot & \cdot \\ \hat{F}_{21} & 1 & \cdot & \cdot & \cdot \\ \hat{F}_{31} & \cdot & \cdot & \cdot & \cdot \end{vmatrix} = m\hat{F}(1-\hat{F})^{m-1} + (1-\hat{F})^m. \tag{61}$$

Let us denote the absolute value of the structure amplitude $|\hat{F}|$ by $\underline{a}$. Then (61), when all the determining signs are positive (which will certainly be so, if all the $\hat{F}$ are positive), has the form

$$D_m^{\oplus} = ma(1-a)^{m-1} + (1-a)^m. \tag{62}$$

But if all the determining signs are negative (and this will occur if all the $\hat{F}$ are negative), then

$$D_m^{\ominus} = -ma(1+a)^{m-1} + (1+a)^m. \tag{63}$$

These functions are given in Tables 6 and 7 for several values of $\underline{m}$.

We see that $D_m^{\oplus}$ decreases monotonically toward 0, remaining positive at all times. The larger the $\underline{m}$, the more rapid is the decrease toward zero.

TABLE 6. Values of $D_m^{\ominus}$

a	$m=5$	$m=10$	$m=20$
0	1	1	1
0.1	0.879	0.236	—5.5043
0.2	0.415	—4.128	—89.454
0.3	— 0.571	—18.028	—687.102
0.4	— 2.305	—53.719	—3944.361
0.5	— 5.063	—134.552	—18843.121
0.6	— 9.175	—302.366	—78580.176
0.7	—15.034	—628.516	—294059.113
0.8	—23.095	—1229.828	—1005694.168
0.9	—33.884	—2291.083	—3185255.572
1.0	—48.0	—4100.0	—9400320.0

TABLE 7. Values of $D_m^{\oplus}$

a	$m=5$	$m=10$	$m=20$
0.1	0.9170	0.736	0.390
0.2	0.740	0.378	0.0690
0.3	0.5300	0.1494	0.007640
0.4	0.3380	0.04645	0.0005238
0.5	0.18745	0.010727	0.000020055

With regard to $D_m^{\ominus}$, its value decreases with increasing rapidity as a increases.

$D_m^{\ominus}$ crosses the axis of abscissas

$$a=\frac{1}{m-1},$$

as is directly apparent from (63).

It is fairly obvious and can be checked by computations, that the curves $D_m = f(a)$ for intermediate cases, i.e., when some of the determining signs are negative, lie between the curves $D_m^{\oplus}$ and $D_m^{\ominus}$.

Therefore, we come to the following important conclusion: If the absolute values of all the $|\hat{F}_{pq}| < \frac{1}{m-1}$, then the determinant is greater than zero for any sign variations and, consequently, the basic inequality does not give any sign information. In other words, the boundary ϵ of the inequality satisfies the condition

$$\varepsilon = \frac{1}{m-1}. \tag{64}$$

Thus, the analysis of the determinant of the third order has meaning only when the structure amplitudes $|\hat{F}| > \frac{1}{2}$, of the fourth order when the $|\hat{F}|$'s are greater than $\frac{1}{3}$, of the fifth order when the $|\hat{F}|$'s are greater than $\frac{1}{4}$, and so forth.

13. The Predominant Positivity of the Structure Product [43]

A connecting determinant D_m of any order is positive. An analysis of the structure of D_m, which may be written as

$$D_m = 1 - \frac{1}{2}\sum_{j\neq k} \hat{F}^2_{jk} + \sum_{i\neq j\neq k} \hat{F}_{ij}\hat{F}_{jk}\hat{F}_{ki} + \ldots,$$

shows that there are reasons to expect predominantly positive structure products. It is true that it is not clear what the influence of the remaining terms of the sum just written will be.

One can assume that, for the given values of $|\hat{F}_{pq}|$, the determinant D_m has the greatest magnitude when all the elements of the determinant, or all the determining signs, are positive.

We shall prove this theorem, having simplified the problem, that is assuming all the $|\hat{F}_{pq}|$ to be identical and equal to a certain $\underline{a}$.

The aspect of the determinant for positive equal $\hat{F}$ was derived in the preceding paragraph:

$$D^{\oplus}_m = ma(1-a)^{m-1} + (1-a)^m.$$

Expanding this according to the binomial formula, we obtain

$$D^{\oplus}_m = 1 - \frac{m(m-1)}{2} a^2 + \sum_{k=3}^{m} (-1)^{k+1} \frac{k-1}{k!} \frac{m!}{(m-k)!} a^k. \qquad (65)$$

This is the expression for D_m when all the determining signs are positive.

If all the determining signs are negative, then the determinant will have the same form as in the case when the signs of all the elements are negative. Consequently (65) is applicable in this case if the signs with $\underline{k}$ odd are changed.

Thus, when all the determining signs are negative,

$$D^{\ominus}_m = 1 - \frac{m(m-1)}{2} a^2 - \sum_{k=3}^{m} \frac{k-1}{k!} \frac{m!}{(m-k)!} a^k. \qquad (66)$$

Let us now go over to the cases when one or several determining signs are negative.

Let S_{23} be negative. Then all the terms of (58) in which a_{23} or a_{32}, but not both, enters as elements, will change sign. D_m will be a polynomial of the same type, but the coefficients will be different. Let us calculate the part of (65) brought in by the terms into which a_{23} enters. Let us emphasize that $a_{23}a_{32}$, but not a_{23} or a_{32} separately enters into the second term of D_m, equal to $\frac{1}{2}$ m (m − 1)a^2.

For the computation of the coefficient of a^k, let us isolate the group of terms

$$a_{23}a_{3\ldots}a_{j\ldots}a_{k\ldots}a_{l\ldots} \;\ldots\; a_{\omega\ldots}.$$

Each addend consists of $\underline{k}$ factors. Of such groups there will be as many as there are methods of choosing k − 2 of the indices j....ω from (m − 2), i.e., from all except the indices 2 and 3. The number of groups is equal to $\frac{(m-2)!}{(m-k)!\,(k-2)!}$. The number of all groups of this type is the same.

With the choice of indices assigned, each group is broken into k − 2 subgroups:

$$a_{23}a_{3j}a_{j}\ldots a_{k}\ldots a_{l}\ldots\ \cdots\ a_{\omega}$$
$$a_{23}a_{3k}a_{j}\ldots a_{k}\ldots a_{l}\ldots\ \cdots\ a_{\omega}$$
$$\cdot\ \cdot\ \cdot\ \cdot\ \cdot\ \cdot\ \cdot\ \cdot\ \cdot\ \cdot\ \cdot$$
$$a_{23}a_{3\omega}a_{j}\ldots a_{k}\ldots a_{l}\ldots\ \cdots\ a_{\omega}$$

The number of these subgroups with identical a is also the same.

A determinant, as is known, does not change if two columns and two rows are interchanged. But the difference between subgroups is exactly this: For example, the first will go over into the second, if the index j is substituted by k, i.e., exchange the places of the jth and kth rows.

There are (k − 2) such subgroups in all. Besides, there are just as many identical subgroups, which may be obtained by the substitution of the second row for the third, and the second column for the third, i.e., the subgroups $a_{2j}a_{32}a_{j\ldots}\ldots a_{\omega\ldots}$ etc.

Thus the terms of each group give

$$2(k-2)\frac{(m-2)\,!}{(m-k)\,!\,(k-2)\,!}\,\gamma_k, \tag{67}$$

where γ_k is the value of the sum of the terms of one subgroup

$$\gamma_k=\sum a_{23}a_{3j}a_{j}\ldots a_{k}\ldots a_{l}\ldots\ \cdots\ a_{\omega}\ldots. \tag{68}$$

The addends in the sum must be formed by permutation of the indices 2,k, l, ..., ω; however, the addends into which at least one cofactor of the type a_{kk}, a_{ll}, ..., $a_{\omega\omega}$ enters must be excluded (since the cofactors are equal to unity, and the number of cofactors a decreases to unity).

Let us prove that

$$\gamma_k=(-1)^{k+1}a^k. \tag{69}$$ [4]

Let us rewrite γ_k in the form

$$\gamma_k=a^2\sum a_{j}\ldots a_{k}\ldots a_{l}\ldots\ \cdots\ a_{\omega}\ldots$$

[4]Equation (69) was derived by V. V. Schmidt.

This can be done, since the cofactor $a_{23}a_{3j}$ in the sum does not change, but the indices 2, k, l, ..., ω are permuted.

Let us now analyze the following determinant:

$$\begin{vmatrix} a_{j2}a_{jk}a_{jl} & \ldots & a_{j\omega} \\ a_{k2}a_{kk}a_{kl} & \ldots & a_{k\omega} \\ a_{l2}a_{lk}a_{ll} & \ldots & a_{l\omega} \\ \cdot\ \cdot\ \cdot\ \cdot & \cdot\ \cdot & \cdot \\ a_{\omega 2}a_{\omega k}a_{\omega l} & \ldots & a_{\omega\omega} \end{vmatrix}.$$

It is obvious that it is equal to $\Sigma a_{j\ldots}a_{k\ldots}a_{l\ldots}a_{\omega\ldots}$ over all the permutations of the indices 2,k,l,... ω. To have all the terms in which there is at least one cofactor of the form a_{kk}, a_{ll},..., $a_{\omega\omega}$ automatically excluded from this sum, let us assume that these elements of the determinant are equal to zero. Then

$$\gamma_k = a^2 \begin{vmatrix} a_{j2} & a_{jk} & a_{jl} & \ldots & a_{j\omega} \\ a_{k2} & 0 & a_{kl} & \ldots & a_{k\omega} \\ a_{l2} & a_{lk} & 0 & \ldots & a_{l\omega} \\ \cdot & \cdot & \cdot & \cdot & \cdot \\ a_{\omega 2} & a_{\omega k} & a_{\omega l} & \ldots & 0 \end{vmatrix} = a^k \begin{vmatrix} 1 & 1 & 1 & \ldots & 1 \\ 1 & 0 & 1 & \ldots & 1 \\ 1 & 1 & 0 & \ldots & 1 \\ \cdot & \cdot & \cdot & \cdot & \cdot \\ 1 & 1 & 1 & \ldots & 0 \end{vmatrix}$$

Let us denote the last determinant of the k − 2 order by D_{k-2}. Let us add to D_{k-2} the determinant

$$D'_{k-2} = \begin{vmatrix} 1 & 1 & 1 & \ldots & 1 \\ 0 & 1 & 0 & \ldots & 0 \\ 1 & 1 & 0 & \ldots & 1 \\ \cdot & \cdot & \cdot & \cdot & \cdot \\ 1 & 1 & 1 & \ldots & 0 \end{vmatrix},$$

which differes from D_{k-2} only in the second row. Evidently

$$D_{k-2} + D'_{k-2} = 0,$$

since in the sum determinant the first and second lines are equal.

But $D'_{k-2} = D_{k-3}$, that is a determinant of the form of D_{k-2}, but of the order k − 3. This is obvious, if D'_{k-2} is developed by the minors of the second row.

Thus, $D_{k-2} + D_{k-3} = 0$, or $D_{k-2} = -D_{k-3}$. But $D_2 = -1$; consequently, $D_3 = 1, \ldots, D_{k-2} = (-1)^{k-1}$, whence

$$\gamma_k = (-1)^{k+1} a^k.$$

If s_{23} is the only negative determining sign, then one should subtract the doubled expression (67) from the coefficients of a^k in (65). Thus, the determinant D_m, when some one determining sign is negative, acquires the form

$$D_m^{1-} = 1 - \frac{m(m-1)}{2} a^2 + \sum_{k=3}^{m} (-1)^{k+1} \times \left[\frac{k-1}{k!} \frac{m!}{(m-k)!} - \frac{4(m-2)!}{(m-k)!(k-3)!} \right] a^k. \quad (70)$$

Table 8 and Fig. 13 give an idea of the magnitude of this function.

For two or three negative determining signs, we shall not be far wrong if we pay no attention to those terms in (65) where these two or three determining signs enter simultaneously. Then

$$D_m^{2-} \approx 1 - \frac{m(m-1)}{2} a^2 + \sum (-1)^{k+1} \times \left[\frac{k-1}{k!} \frac{m!}{(m-k)!} - \frac{8(m-2)!}{(m-k)!(k-3)!} \right] a^k, \quad (71)$$

$$D_m^{3-} \approx 1 - \frac{m(m-1)}{2} a^2 + \sum (-1)^{k+1} \times \left[\frac{k-1}{k!} \frac{m!}{(m-k)!} - \frac{12(m-2)!}{(m-k)!(k-3)!} \right] a^k. \quad (72)$$

Figure 14 shows graphs for $D_m^{\oplus}$, $D_m^{\ominus}$, D_m^{1-}, D_m^{2-} when m = 10.

The result of this computation, even when it is realized that it cannot be fully applied to a real case in which the elements $|\hat{F}_{pq}|$ are not equal to one another, shows quite clearly that the overwhelming majority of the strong structure products must be positive. This effect is the greater the greater the absolute values of the $|\hat{F}|$ that figure in D_m. Since the fraction of strong $|\hat{F}|$'s increases as the number of atoms in the cell decreases, it may be said that the predominant positivity of the structure products is revealed the more clearly, the smaller the number of atoms in the unit cell.

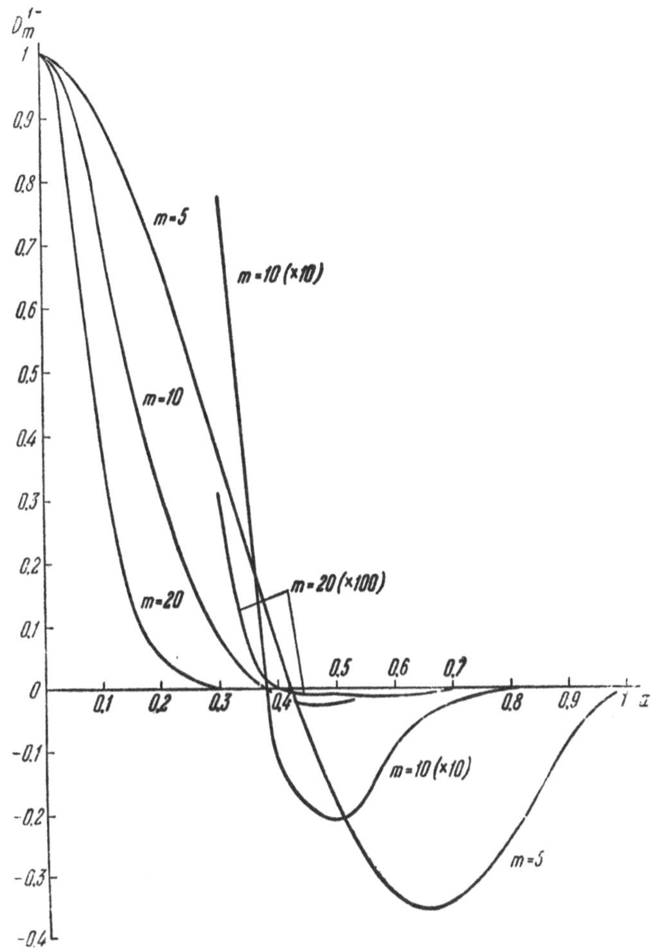

Fig. 13. Curves of D_m^{1-} as a function of a.

The limiting values of a increases very rapidly with an increase in the number of negative determining signs. We see, for example, that $D_{10}^{1-} > 0$ for amplitudes greater than 0.38.

Of 36 determining signs only one can be negative if the amplitudes are greater than 0.31; only two signs can be negative if the amplitudes are greater than 0.27, etc.

Thus, the connecting equation of the amplitudes leads, in principle, to the possibility of picking out positive structure products.

TABLE 8

Values of D_m^{1-}				Roots of D_m^{1-}	
a	D_m^{1-}			m	a
	$m=5$	$m=10$	$m=20$		
0	1	1	1	3	0.50
0.1	0.9008	0.7209	0.3798	4	0.45
0.2	0.6758	0.3221	0.0562	5	0.42
0.3	0.3695	0.0781	0.0031	6	0.41
0.4	0.0605	—0.0110	—0	7	0.39
0.5	—0.1875	—0.0205	—0.0001	8	0.39
0.6	—0.3277	—0.0096	—0.0002	9	0.38
0.7	—0.3397	—0.0023	0.0000	10	0.38
0.8	—0.2390	—0.0003	0.0000	11	0.37
0.9	—0.0870	—0.0001	0.0000	12	0.37
1.0	0	0	0	13	0.37
				14	0.36
				15	0.36
				16	0.35
				17	0.34
				18	0.34
				19	0.34
				20	0.33

TABLE 9. Values of D_{10}^{2-}

a	D_{10}	a	D_{10}
0	1	0.6	—0.0211
0.1	0.7054	0.7	—0.0044
0.2	0.2684	0.8	—0.0004
0.3	0.0046	0.9	—0.0002
0.4	—0.0633	1.0	0
0.5	—0.0518		

14. Finding Positive Structure Products

The basic result of the computation just performed is a table of roots of D^{1-}_m.

According to the fundamental theorem, $D_m \geq 0$, i.e., a determinant of any order cannot be negative. If therefore, for amplitudes of greater $\underline{a}$, it were possible to form a determinant of the $\underline{m}$th order D^{1-}_m (Table 8, page 170), then all the determining signs in this determinant are reliably positive.

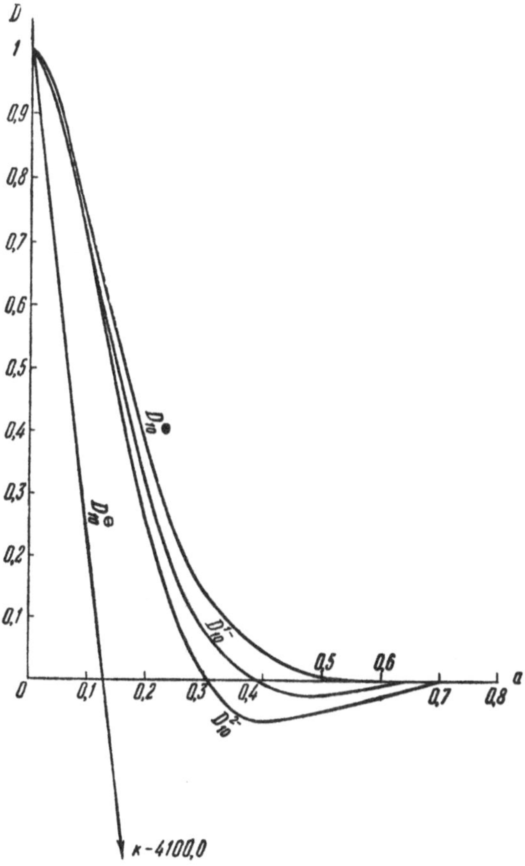

Fig. 14. The different curves of D_{10} as a function of $\underline{a}$.

It follows from the tables and graphs that the limiting values converge strongly when two or more determining signs are negative. If a determinant of the order $\underline{m}$ is formed for amplitudes with greater $\underline{a}$, for which $D_{\underline{m}}^{1-} > 0$, but $D_{m}^{2-} < 0$, then it means that one (no more than one) of the determining signs cannot be negative. Analogously, if $D_{m}^{2-} > 0$, but $D_{m}^{3-} < 0$, then two determining signs can be negative.

Thus, the number of structure products that can be negative, of the total that enter into the determinant, can be reliably established.

However, with the impossibility of forming $D_{m}^{1-} < 0$, we are also deprived of the possibility of reliably isolating positive structure products. In this case the law of the predominantly positive sign of structure products cannot be applied in practice without having recourse to probability considerations (see Sections 19 and 20).

It is essential to note that the condition $D_{m}^{2-} < 0$ indicates the absence of negative signs for special determinants among whose elements occur amplitudes related by symmetry.

Let us analyze, for example, the following seventh order determinant

$$\begin{vmatrix} 1 & \cdots & \cdots & \cdots & \cdots & \cdots & \cdots \\ H_1 & 1 & \cdots & \cdots & \cdots & \cdots & \cdots \\ H_2 & H_2-H_1 & 1 & \cdots & \cdots & \cdots & \cdots \\ H_3 & H_3-H_1 & H_3-H_2 & 1 & \cdots & \cdots & \cdots \\ H_2+H_1 & H_2 & H_1 & H_2+H_1-H_3 & 1 & \cdots & \cdots \\ H_1+H_3 & H_3 & H_1-H_2+H_3 & H_1 & H_3-H_2 & 1 & \cdots \\ H_3+H_2 & H_3+H_2-H_1 & H_3 & H_2 & H_3-H_1 & H_2-H_1 & 1 \end{vmatrix} \qquad (73)$$

Keeping in mind that special determinants of this type may be formed for different $\underline{m}$, we take into account the condition that $D_{m}^{2-} < 0$, as well as the limiting criteria of reliable judgements concerning the signs of the structure products.

Let us analyze the possibilities in principle of finding positive structure products.

Let us assume that the field of indices contains 10,000 structure amplitudes distributed according to the Gaussian law (see Chapter III). The determinant composed of the strongest amplitudes has the greatest capabilities. Let us analyze the optimal case – the determinant is formed from the strongest amplitudes. The mean value of $|\hat{F}|$ is taken as $\underline{a}$. To form D_5, it is necessary to have 10 amplitudes; for D_{10}, 45 amplitudes; for D_{20}, 190 amplitudes; and D_{40}, 780 amplitudes.

If the root mean square is $\sqrt{\overline{F^2}} = \sigma$, then five out of the 10,000 reflections will be stronger than 3.9σ, 45 of 10,000 stronger than 2.856σ, 190 of 10,000 stronger than 2.34σ, and 780 of 10,000 stronger than 1.76σ.

Whence we obtain the following largest possible values of $\underline{a}$ for the determinant D_m (the last four columns of Table 10):

TABLE 10. Values of $\underline{a}$

m	Roots of D_m^{1-}	Roots of D_m^{2-}	No. of amplitudes entering into D_m	$\sigma = 0.1$ ($N = 100$)	$\sigma = 0.12$ ($N = 70$)	$\sigma = 0.15$ ($N = 45$)	$\sigma = 0.20$ ($N = 25$)
5	0.42		10	0.40	0.47	0.58	0.78
10	0.33	0.31	45	0.28	0.34	0.41	0.57
20	0.33		190	0.23	0.28	0.35	0.47
40	0.30		780	0.18	0.21	0.26	0.35

The values of $\underline{a}$ are given for various σ. The number N of effective atoms in the structure that corresponds to the given σ is indicated in parentheses. With 100 atoms in the cell, the greatest amplitudes have a magnitude of about 0.4, with 25 atoms about 0.8.

In the second column of Table 10 are given the roots of D_m^{1-}, i.e., the magnitudes $\underline{a}$ at which $D_m^{-1} = 0$. Consequently, D_m^{1-} constructed for a larger $\underline{a}$, contains reliably positive structure amplitudes. Comparing the figures of this column with those in the last four columns, we come to the following conclusion.

When N = 25, all the magnitudes of this column are larger than the corresponding magnitudes of the second column. Therefore, in a structure with 25 effective atoms it is possible, in principle, to construct in practice determinants of all orders for values of $\underline{a}$ larger than the limiting ones. These will contain only reliably positive structure products. Thus, direct structure analysis will always be possible for such structures.

The same refers also to structures with $\sigma = 0.15$; here the roots of D_m^{1-} are smaller than the values of $\underline{a}$ actually attainable for m = 5, 10, and 20. The determinants of these orders may in principle be constructed and allow the selection of positive structure products.

With a further increase of the number of atoms the chances of applying D_m^{1-} for the finding of reliably positive signs worsens. When N = 100 determinants of even the lowest orders already may be formed only for values of $\underline{a}$ smaller than the limiting ones.

If we have succeeded in forming a special determinant into whose composition of original amplitudes enter $\mathbf{H}_i$, $\mathbf{K}$ and $\mathbf{H}_i + \mathbf{K}_i$ H_i, K_i and $H + K_i$, then (as has been explained on page 172) the determinant "degenerates" and the role of D_m^{1-} is taken over by D_m^{2-}, since all the structure products are "twinned."

The values of D_m^{2-} are given in the third column of the table. We see that the use of such a determinant increases the possibilities of selecting reliably positive structure products. If the application of D_{10}^{1-} for this purpose becomes impossible for $\sigma < 0.14$, then the application of D_{10}^{2-} already becomes impossible at $\sigma < 0.11$.

This analysis has a basic significance for structural studies, as it indicates a precise limit for the reliable determination of the signs of structure products. This limit is determined by the complexity of the structure and lies at $\sigma = 0.10$ to 0.15.

It must be emphasized that this is not at all a disappointing result. In the case of identical atoms their number in the cell, 50-100, may appear negligible. However, one must remember that σ increases appreciably with the presence of a heavy atom in the structure.

It is not difficult to be convinced that the overwhelming majority of structure determinations carried out at the present time is for substances with σ larger than the boundary indicated. Thus, in modern structure analysis, methods of reliably determining signs may be widely used. This is illustrated in Table 11.

TABLE 11. The Magnitude of σ for Some Compounds

Compound	Number of molecules in noncentered cells	σ
$K_3UO_2F_5$	4	0.24
p-aminosalycylic acid	4	0.16
WAl_{12}	2	0.26
$C_6H_5SeO_2H$	4	0.21
$C_{14}H_{29}COOH$	4	0.11

Only for the analysis of the last compound is it impossible to apply fully reliable correlations between amplitudes for the determination of signs.

In the preceding chapter we have discussed the probability $\underline{w}$ of a positive sign of a structure product. The theory of the connecting determinant indicates the presence of a strong correlation between structure products for large $\hat{F}$. There is a strong correlation when the structure products are larger than $4\text{-}5\sigma^3$ (see Chapter III). For identical atoms in the cell this means the following: When N = 100 the magnitude of $\sigma = 0.1$ and $\sqrt{5\sigma^3} = 0.17$. Consequently, the correlation is already substantial when the structure amplitudes that form a structure product are larger than 0.2. When N = 25 the magnitude of $\sigma = 0.2$ and $\sqrt{5\sigma^3} = 0.34$. The correlation is substantial for amplitudes larger than 0.35.

For structure products formed from $|\hat{F}|$ greater than the limits just given, the application of reliable methods for sign determination becomes possible – which we see from Table 10. Simultaneously with this, the strong correlation that arises between structure products does not allow one to carry out the computation of W_+ for this product independently of the magnitudes of the remaining ones. Thus, it follows from the theory of the connecting determinant that for large products the probability concept $W_+(x)$ loses its meaning (see Section 20). Constructing D_m^{n-}, it is possible to say which number $\underline{n}$ of the given $M = (m-1)(m-2)$ of structure products can be negative. The fraction $\dfrac{n}{M}$ to some extent characterizes the probability $(1 - W_+)$ for any structure product that enters into a given connecting determinant.

The formulas and graphs given for D_m^{1-} and D_m^{2-} show that the probabilities W_+ of large structure products, even for complex structures, substantially exceed 0.5.

See page 201 for further discussion of this important question.

15. Relations Between the Absolute Magnitudes of Structure Amplitudes

For simplicity let us divide all the amplitudes that enter into a determinant into two groups: Basic, whose magnitude we shall assume to be $\underline{a}$ and others with a magnitude of $\underline{x}$.

Let us consider the determinant

$$D_m^{ax} = \begin{vmatrix} 1 & a & a & a & \dots & a \\ a & 1 & x & x & \dots & x \\ a & x & 1 & x & \dots & x \\ \cdot & \cdot & \cdot & \cdot & \cdot & \cdot \\ a & x & x & x & \dots & 1 \end{vmatrix}.$$

We shall proceed from the theorem which was proved for the particular case of equal elements in the preceding paragraph. The determinant D_m has the greatest value if all its elements are greater than zero.

If all the a and all the x are positive, then D_m has the greatest value. On the other hand, according to the fundamental theorem of connection $D_m \geq 0$.

Let a be fixed, and x decrease. Then, as will be shown presently, at some $x = x_0$ the determinant D_m will become zero and thereafter will become negative. This is impossible; therefore, for a given a the value of x cannot be smaller than x_0 if all the signs of $\hat{F}_{pq}$ are positive. Even more strongly, x cannot be smaller than x_0 if among the $\hat{F}_{pq}$ some sign are negative – in this case D_m will be still smaller. Consequently, $x_0 = f(a)$ (the root of the equation $D_m^{ax} = 0$) is the minimum value of x for a given a independently of the signs of the $\hat{F}_{pq}$.

In the example of the simplest inequality $\hat{F}_H^2 \leq \frac{1}{2} + \frac{1}{2} \hat{F}_{2H}$, we saw that next to the field of action of the inequality is a forbidden region. The finding of this forbidden area in the general case for different $|\hat{F}_{pq}|$ has been analyzed in Sections (17) and (18). But here our goal is to characterize roughly by a linear function the region forbidden by the general inequality itself $D_m \geq 0$.

Let us find $x_0 = f(a)$. Expressing D_m^{ax} as a sequence of minors and applying (62) for D_m with equal elements, we obtain

$$D_m^{ax} = (m-1)(1-x)^{m-2}(x-a^2) + (1-x)^{m-1}. \qquad (74)$$

The unique root of D^{ax} is expressed by the simple formula

$$x_0 = \frac{(m-1)a^2 - 1}{m-2}. \qquad (75)$$

A limitation on the possible values of x appears beginning with $a = \frac{1}{\sqrt{m-1}}$, and as a increases further, an increasingly greater part of small x is forbidden.

For the cases m = 3 and m = 4,

$$x_0 = 2a^2 - 1 \text{ and } x_0 = \frac{1}{2}(3a^2 - 1).$$

It is always possible to evaluate the limiting region for each particular D_m. However, this is not necessary, since only the general laws are relevant.

For example, let us take nine original structure amplitudes $\hat{F}_{\mathbf{H}_i}$ greater than $\underline{a}$ and construct the determinant D_{10}, into which enter the amplitudes $|\hat{F}_{\mathbf{H}_i - \mathbf{H}_k}|$. The absolute value of these amplitudes is limited; it is greater than x_0, i.e., greater than $\frac{9a^2-1}{8}$. Therefore, if nine original amplitudes $|\hat{F}_{\mathbf{H}_i}| > \frac{1}{2}$, then $|\hat{F}_{\mathbf{H}_i - \mathbf{H}_k}| > \frac{5}{32}$.

Thus, one can ascertain the presence of the following connection between structure amplitudes. If there are (m − 1) amplitudes $\hat{F}_{\mathbf{H}_i}$ the absolute values of which are greater than $\underline{a}$, then the amplitudes of $\hat{F}_{\mathbf{H}_i \pm \mathbf{H}_k}$ are greater in absolute magnitude than $x_0 = \frac{(m-1)a^2 - 1}{m-2}$.

An anthracene crystal has eleven amplitudes greater than $\frac{1}{2}$. Consequently, the amplitudes $\hat{F}_{\mathbf{H} \pm \mathbf{K}}$ constructed by combining these eleven indices, must be in absolute magnitude greater than $\frac{7}{40} = 0.175$. Of all the possible 110 combinations $\mathbf{H} + \mathbf{K}$, only three or four combinations disobey this rule, but still give $|\hat{F}| \approx 0.1$.

A very important result of this paragraph is the proof that in the situation when $\hat{F}_{\mathbf{H}}$ and $\hat{F}_{\mathbf{K}}$ are strong, the magnitudes of $\hat{F}_{\mathbf{H} \pm \mathbf{K}}$ are also predominantly strong.

We are able to determine, although roughly, on the basis of considerations of the boundary values of $\hat{F}_{\mathbf{H} \pm \mathbf{K}}$, the fraction of strong structure products as it varies with the structure.

Let us assume that we have at our disposal (m − 1) strong structure amplitudes, the absolute magnitudes of which are greater than $\underline{a}$. It is clear that the greater the $\underline{m}$, the smaller the $\underline{a}$, since the magnitudes of the amplitudes are distributed according to the Gaussian law. We are interested in the magnitudes of the remaining structure amplitudes which will enter into the determinant D_m.

We shall obtain an approximate conception of this by applying (75). From the $\underline{a}$ and $\underline{m}$ assigned we can find the magnitude of x_0 from this equation and we may assume that in the determinant D_m with $(m-1)$ supporting amplitudes greater than $\underline{a}$, the values of the remaining amplitudes will be greater than x_0. The amplitudes that entered in D_m will give $\frac{(m-1)(m-2)}{2}$ structure products greater than a^2x_0.

For a given $\underline{m}$ the magnitude $\underline{a}$ and, consequently x_0 depends only on σ, the root mean square value of the structure amplitude. Therefore, with a given σ we can always find the number of structure products a^2x_0 greater than the value that interests us, that is it is possible to form $\frac{(m-1)(m-2)}{2}$ structure products greater than $a^2\frac{(m-1)a^2-1}{m-2}$, where $\underline{a}$ is the abcissa of the Gaussian curve taken for the ordinate $\frac{m-1}{M}$, if M is the total number of measured reflections.

If the number of reflections is 5000, we find according to the Gaussian curve (Table 12).

TABLE 12

$(m-1)$	5	10	20	40	80
a	3.33σ	3σ	2.88σ	2.65σ	2.41σ
x_0	$\frac{5}{4}\,11.1\sigma^2-\frac{1}{4}$	$\frac{10}{9}\cdot 9\sigma^2-\frac{1}{9}$	$\frac{20}{19}\cdot 8.3\sigma^2-\frac{1}{19}$	$\frac{40}{39}\cdot 7\sigma^2-\frac{1}{39}$	$\frac{80}{79}\cdot 5.8\sigma^2-\frac{1}{79}$

For example, for m = 41, 40 amplitudes will be found stronger than $a = 2.65\sigma$, and, consequently, the values of the remaining amplitudes that entered into D_m will be greater than $\frac{40}{39}\,7\sigma^2-\frac{1}{39}$. Thus, for $\sigma = 0.2$, we obtain approximately 800 structure products larger than 0.06.

Let us note, moreover, that this makes obvious the ease with which structures of such a degree of complexity may be solved (20-30 identical atoms in the cell). From Fig. 20 (page 202) it is clear that an overwhelming majority of the structure products greater than 0.06 will be reliably positive.

The considerations stated in this paragraph are of an orientational character. They are necessary if the researcher wishes to obtain general a priori conclusions. As concerns the connections between the magnitudes and signs of particular structure amplitudes, they can be found with the aid of the formulas to be studied below (see Section 17 and those following).

16. Signs of Structure Products and Amplitudes

We have already noted that with the aid of inequalities, the signs of structure products and not those of structure amplitudes are determined.

This completely general statement has the following meaning. In a centrosymmetric crystal the origin of coordinates may be selected in different ways. With a change in the selection of the origin of coordinates the signs of a series of structure amplitudes may be reversed.

The sign of a structure product does not depend on the choice of the origin of coordinates. This is quite obvious from the general formula for a structure product in this notation:

$$X_{\mathbf{H},\mathbf{K},\mathbf{L}\ldots}=\sum_{j_1}\sum_{j_2}\sum_{j_3}\ldots n_{j_1}n_{j_2}n_{j_3}\ldots\exp 2\pi i\,[\mathbf{H}\mathbf{r}_{j_1}+\mathbf{K}\mathbf{r}_{j_2}+ \\ +\mathbf{L}\mathbf{r}_{j_3}+\ldots-(\mathbf{H}+\mathbf{K}+\mathbf{L}+\ldots)\,\mathbf{r}_j], \tag{76}$$

or

$$X_{\mathbf{H},\mathbf{K},\mathbf{L}\ldots}=\sum n_{j_1}n_{j_2}n_{j_3}\ldots\exp 2\pi i\,[\mathbf{H}(\mathbf{r}_{j_1}-\mathbf{r}_j)+ \\ +\mathbf{K}(\mathbf{r}_{j_2}-\mathbf{r}_j)+\mathbf{L}(\mathbf{r}_{j_3}-\mathbf{r}_j)+\ldots].$$

This means that a structure product is determined by the magnitudes of interatomic vectors and, consequently, is unaffected by the choice of the origin of coordinates.

Furthermore, it follows from this condition that the signs of structure amplitudes with even indices are independent of the choice of the origin of coordinates. Actually, from the definition of the structure product, it is clear that the signs of degenerate structure products and those of even structure amplitudes coincide.

Thus, the sign of a structure product of any order is determined by the interatomic vectors and is independent of the choice of the origin of coordinates of the cell at one or another of its centers of inversion.

On the other hand, it is clear that by transferring to another center of inversion, i.e., shifting one or several coordinates by $\frac{1}{2}$ or $\frac{1}{4}$, we can change the sign of a structure amplitude.

Evidently, in agreement with the three possible shifts of the origin of coordinates, it is always possible to choose arbitrarily the signs of three amplitudes H_1, H_2 and H_3. However, if the sign of one amplitude is chosen, then the choice of signs of the remaining amplitudes must occur according to definite rules.

The essence of these rules comes to the following: Perhaps the sign of that $\hat{F}_H$ is chosen arbitrarily which does not follow uniquely from the choice of sign of the structure product.

The Choice of the First Sign. Only degenerate products of the type $\hat{F}^2_H\hat{F}_{2H}$ or $\hat{F}^2_H\hat{F}^2_K\hat{F}_{2H+2K}$ determine by their sign the sign of $\hat{F}_{2H}$. From this it follows that the first sign may be arbitrarily assigned to any index hkl which has at least one of the three numbers odd. However, if there is an even index $2H$ in centered groups, but extinctions transform $\hat{F}_H$ into zero, then a sign may also be assigned arbitrarily to such an amplitude [for example, the index hkl (4, 6, 16) is suitable, if centering extinguishes the reflections $h+k \neq 2n$]. Thus, in a more general form: The first sign may be assigned to any amplitude $\hat{F}_{H_1}$, only if there is no amplitude $\hat{F}_{\frac{1}{2}H_1}$ different from zero.

The Choice of the Second Sign. Let us examine the structure product $\hat{F}^2_H\hat{F}_K\hat{F}_{2H+K}$.

We see that, for a given sign of a structure product, the sign $\hat{F}_K$ determines the sign of $\hat{F}_{2H+K}$. Consequently, the second sign must be chosen in such a way that the difference of the first and second indices does not give an even index.

Again, the given product exists only if there is an amplitude $\hat{F}_H$ different from zero. Thus, in centered lattices a second index may be chosen so that the difference of indices equals $2H$ if the amplitude $\hat{F}_H$ is extinguished.

Thus, the general rule is this: The second sign may be assigned arbitrarily to any amplitude $\hat{F}_{H_2}$, only if there are no amplitudes $\hat{F}_{\frac{1}{2}H_2}$ and $\hat{F}_{\frac{1}{2}(H_2-H_1)}$ different from zero.

The Choice of the Third Sign. Here a new prohibition is superimposed on the product $\hat{F}_H\hat{F}_K\hat{F}_{H+K}$. Evidently the third amplitude, the sign of which is established arbitrarily, may have any index $\mathbf{H}_3 \neq \mathbf{H}_2 \pm \mathbf{H}_1$, only if there are no amplitudes $\hat{F}_{\frac{1}{2}\mathbf{H}_3}$, $\hat{F}_{\frac{1}{2}(\mathbf{H}_3-\mathbf{H}_1)}$, $\hat{F}_{\frac{1}{2}(\mathbf{H}_3-\mathbf{H}_2)}$ different from zero.

Example:

The Primitive Cell

The first index: 548 is suitable, 648 is not.

The second index: 862 is unsuitable $\left(F_{\frac{1}{2}\mathbf{H}_2} \neq 0\right)$, 7 6 10 is unsuitable $\left(F_{\frac{1}{2}(\mathbf{H}_3-\mathbf{H}_1)}\right) \neq 0$, 372 is suitable.

The third index: 426 is unsuitable $\left(F_{\frac{1}{2}\mathbf{H}_3} \neq 0\right)$, 748 is unsuitable $\left(F_{\frac{1}{2}(\mathbf{H}_3-\mathbf{H}_1)} \neq 0\right)$, 752 is unsuitable $\left(F_{\frac{1}{2}(\mathbf{H}_3-\mathbf{H}_2)} \neq 0\right)$, 431 is suitable.

A Centered Cell

$h + k = 2n + 1$ are absent.

The first index: 512 is suitable, 428 is suitable (!)$\left(\text{since } F_{\frac{1}{2}\mathbf{H}_1} = 0\right)$, 228 is unsuitable $\left(F_{\frac{1}{2}\mathbf{H}_1} \neq 0\right)$.

The second index: 862 is suitable, 662 is unsuitable, etc.

17. The Limits of Possible Values of Structure Amplitudes

Since the connecting determinant has a value between 0 and + 1, and moreover, as shown in Table 6, is usually close to zero, all but one of the amplitudes $\hat{F}_{pq}$ which enter into the determinant can have quite arbitrary values; the value of this last one cannot be arbitrary, but lies between certain limits, sometimes rather narrow ones.

Let us set ourselves the task of finding these limits [40]. Let us study the structure amplitudes $\hat{F}_{pq}$, which compose a connecting determinant of order $\underline{m}$. Let us assume the number of atoms N in the cell to be greater than $\underline{m}$. Then $D_m > 0$ and, consequently, the vectors $\mathbf{G}_p$ will be linearly independent (compare page 153).

It is not hard to see that we have a right to choose the orthonormal vector system $\mathbf{e}_j$ of the N-dimensional space in such a way that

$$\mathbf{G}_p = \sum_{j=1}^{p} \alpha_{pj}\mathbf{e}_j. \tag{77}$$

In fact, we may choose the first vector so that it is parallel to $\mathbf{G}_1$, i.e., set

$$\mathbf{G}_1 = \mathbf{e}_1,$$

since $\alpha_{11} = 1$. The second vector can be drawn so that

$$\mathbf{G}_2 = \alpha_{21}\mathbf{e}_1 + \alpha_{22}\mathbf{e}_2.$$

The third vector drawn so that

$$\mathbf{G}_3 = \alpha_{31}\mathbf{e}_1 + \alpha_{32}\mathbf{e}_2 + \alpha_{33}\mathbf{e}_3,$$

and so forth, and this is what gives (77).

Let all the structure amplitudes $\hat{F}_{pq}$ that enter into the Gram determinant of the vectors $\mathbf{G}_p$ be known, except one, let us say $\hat{F}_{m,m-1}$. This means that the cosines of all the angles between the N-dimensional vectors are fixed, except the angle between the vectors $\mathbf{G}_m$ and $\mathbf{G}_{m-1}$. Let us imagine that $(m-1)$ of the vectors $\mathbf{G}_p$ $(p = 1,\ldots,m-1)$ are constructed, and let us see what freedom there remains for vector $\mathbf{G}_m$. Since the cosines of the angles of $\mathbf{G}_m$ with the vectors $\mathbf{G}_p$ $(p = 1\ldots m-2)$ are fixed, the tip of the vector $\mathbf{G}_m$ lies in a two-dimensional plane determined by the system of equations

$$(\mathbf{x}\mathbf{G}_p) = \hat{F}_{mp} \quad (p = 1\ldots m-2). \tag{78}$$

But the tip of $\mathbf{G}_m$ lies, in addition, on the surface of an m-dimensional sphere and describes the circumference cut out on the sphere by the plane (78).

The components of $\mathbf{G}_m$ are fixed by their components, except $\mathbf{e}_{m-1}$ and $\mathbf{e}_m$. The plane (78) is parallel to the plane $\mathbf{e}_{m-1}\mathbf{e}_m$. The projection of $\mathbf{G}_m$ on this plane is equal to, in accordance with (77),

$$\alpha_{m,m-1}\mathbf{e}_{m-1} + \alpha_{m,m}\mathbf{e}_m.$$

Introducing the azimuthal angle $\tan\varphi = \alpha_{m,m}/\alpha_{m,m-1}$ we shall be able to express $\hat{F}_{m,m-1}$ in terms of the remaining $\hat{F}_{pq}$ and the single independent parameter φ.

To do this α_{pq} has to be expressed in terms of $\hat{F}_{pq}$, by forming scalar products between the vectors (77). Doing this successively up to p = m − 2, we obtain

$$\left.\begin{aligned} &\hat{F}_{p1} = \alpha_{p1};\\ &\hat{F}_{p2} = \alpha_{p1}\alpha_{21} + \alpha_{p2}\alpha_{22};\\ &\hat{F}_{p3} = \alpha_{p1}\alpha_{31} + \alpha_{p2}\alpha_{32} + \alpha_{p3}\alpha_{33};\\ &\hat{F}_{pq} = \sum_{1}^{q} \alpha_{pi}\alpha_{qi}. \end{aligned}\right\} \tag{79}$$

Since the system of basic vectors is orthogonal, then

$$\left.\begin{aligned} &\alpha_{22}^2 = 1 - \alpha_{21}^2; \quad \alpha_{33}^2 = 1 - \alpha_{31}^2 - \alpha_{32}^2, \ldots;\\ &\alpha_{pp}^2 = 1 - \sum_{1}^{p-1} \alpha_{pi}^2. \end{aligned}\right. \tag{80}$$

With the aid of these relations we obtain the values of all the α, up to $\alpha_{m,m-2}$.

Taking into account that

$$\alpha_{m,m}^2 + \alpha_{m,m-1}^2 = 1 - \alpha_{m1}^2 - \alpha_{m2}^2 - \ldots - \alpha_{m,m-2}^2$$

and using

$$\alpha_{m,m}^2 + \alpha_{m,m-1}^2 = \alpha_{m,m-1}^2/\cos^2\varphi, \tag{81}$$

we are able to express

$$\hat{F}_{m,m-1} = \sum_{i=1}^{m-1} \alpha_{mi}\alpha_{m-1,i} \tag{82}$$

in terms of the remaining $\hat{F}_{pq}$ and the azimuthal angle φ.

Let us perform this operation for determinants of low order.

1) m = 3, $\hat{F}_{m,m-1} = \hat{F}_{32}$.

According to (82)

$$\hat{F}_{32} = \alpha_{31}\alpha_{21} + \alpha_{32}\alpha_{22}.$$

According to (79)

$$\alpha_{31} = \hat{F}_{31}; \quad \alpha_{21} = \hat{F}_{21}.$$

and to (81)

$$\frac{\alpha_{32}^2}{\cos^2\varphi} = 1 - \hat{F}_{31}^2.$$

According to (80)

$$\alpha_{22}^2 = 1 - \hat{F}_{21}^2.$$

Consequently,

$$\hat{F}_{32} = \hat{F}_{21}\hat{F}_{31} + \sqrt{\left(1 - \hat{F}_{21}^2\right)\left(1 - \hat{F}_{31}^2\right)} \cos\varphi. \tag{83}$$

2) $m = 4$, $\hat{F}_{m,m-1} = \hat{F}_{43}$,

According to (82)

$$\hat{F}_{43} = \alpha_{41}\alpha_{31} + \alpha_{42}\alpha_{32} + \alpha_{43}\alpha_{33}.$$

and to (79)

$$\hat{F}_{41} = \alpha_{41}; \quad \hat{F}_{31} = \alpha_{31}; \quad \hat{F}_{42} = \alpha_{41}\alpha_{21} + \alpha_{42}\alpha_{22}.$$
$$\hat{F}_{32} = \alpha_{31}\alpha_{21} + \alpha_{32}\alpha_{22}.$$

According to (80)

$$\alpha_{22} = \sqrt{1 - \alpha_{21}^2}.$$

whence

$$\alpha_{32} = \frac{1}{\sqrt{1 - \hat{F}_{21}^2}} (\hat{F}_{32} - \hat{F}_{31}\hat{F}_{21}); \quad \alpha_{42} = \frac{1}{\sqrt{1 - \hat{F}_{21}^2}} (\hat{F}_{42} - \hat{F}_{41}\hat{F}_{21}).$$

According to (81)

$$\frac{\alpha_{43}^2}{\cos^2\varphi} = 1 - \alpha_{41}^2 - \alpha_{42}^2.$$

and to (80)

$$\alpha_{33}^2 = 1 - \alpha_{31}^2 - \alpha_{32}^2.$$

Substituting the values of α_{42} and α_{32}, we obtain

$$\left.\begin{aligned} \alpha_{43} &= \sqrt{1 - \hat{F}_{41}^2 - \frac{1}{1-\hat{F}_{21}^2}(\hat{F}_{42} - \hat{F}_{41}\hat{F}_{21})^2}\cos\varphi; \\ \alpha_{33} &= \sqrt{1 - \hat{F}_{31}^2 - \frac{1}{1-\hat{F}_{21}^2}(\hat{F}_{32} - \hat{F}_{31}\hat{F}_{21})^2}. \end{aligned}\right\} \tag{84}$$

Thus

$$\hat{F}_{43} = \hat{F}_{41}\hat{F}_{31} + \frac{1}{1-\hat{F}_{21}^2}(\hat{F}_{42} - \hat{F}_{41}\hat{F}_{21})(\hat{F}_{32} - \hat{F}_{31}\hat{F}_{21}) +$$
$$+ \sqrt{\left[1 - \hat{F}_{41}^2 - \frac{1}{1-\hat{F}_{21}^2}(\hat{F}_{42} - \hat{F}_{41}\hat{F}_{21})^2\right]\left[1 - \hat{F}_{31}^2 - \frac{1}{1-\hat{F}_{21}^2}(\hat{F}_{32} - \hat{F}_{31}\hat{F}_{21})^2\right]}\cos\varphi. \tag{85}$$

When necessary it is not difficult only cumbersome, to derive the corresponding formulas for greater values of m.

We see that in all the cases

$$\hat{F}_{m,\, m-1} = A + B\cos\varphi, \tag{86}$$

where A and B are expressed in terms of the remaining $\hat{F}_{pq}$.

The limiting values sought are

$$A - B \leqslant \hat{F}_{m,\, m-1} \leqslant A + B. \tag{87}$$

It is quite clear, that the inequalities $D_m \geq 0$ are a consequence of (86). Squaring this, we obtain

$$(\hat{F}_{m,\, m-1} - A)^2 = B^2\cos^2\varphi,$$

i.e.,

$$(\hat{F}_{m,\, m-1} - A)^2 \leqslant B^2. \tag{88}$$

For m = 3, squaring (83), we have

$$\left(\hat{F}_{32}-\hat{F}_{21}\hat{F}_{31}\right)^2 \leqslant \left(1-\hat{F}^2_{21}\right)\left(1-\hat{F}^2_{31}\right).$$

Denoting $\hat{F}_{21} = \hat{F}_{\mathbf{H}}$, $\hat{F}_{31} = \hat{F}_{\mathbf{K}}$ (then $\hat{F}_{32} = \hat{F}_{\mathbf{K}-\mathbf{H}}$), we see that the last inequality is nothing but (52). Likewise, squaring (85), we obtain the inequality (54), etc.

A knowledge of the limiting values of a structure amplitude may be of substantial value in determining signs of amplitudes or in performing some other analysis of the experiment. A computation according to (83) (for $\cos\varphi =$ $= \pm 1$) gives Table 13.

The limiting values of $\hat{F}_{\mathbf{K}-\mathbf{H}}$ are given in the squares of Table 13. If the sign of $\hat{F}_{\mathbf{H}}\hat{F}_{\mathbf{K}}$ is negative, then the signs given in the table should be reversed.

It is obvious from Table 13 which values of $|\hat{F}_{\mathbf{H}}|$ and $|\hat{F}_{\mathbf{K}}|$ determine the sign of $\hat{F}_{\mathbf{K}-\mathbf{H}}$ unequivocally. This part of the table is marked off by a thick line. Analytically this condition of unequivocal determination of the sign of $\hat{F}_{\mathbf{K}-\mathbf{H}}$ for certain values of $|\hat{F}_{\mathbf{H}}|$ and $|\hat{F}_{\mathbf{K}}|$ can be written thus

$$|\hat{F}_{\mathrm{H}}\hat{F}_{\mathrm{K}}| - \sqrt{\left(1-\hat{F}^2_{\mathrm{H}}\right)\left(1-\hat{F}^2_{\mathrm{K}}\right)} \geqslant 0$$

or

$$1-\hat{F}^2_{\mathrm{H}}-\hat{F}^2_{\mathrm{K}} \leqslant 0. \tag{89}$$

Equation (83) and its consequences are correct irrespective of what values the other structure amplitudes have (only three amplitudes enter into a determinant of order m = 3). The higher the order of the determinant, the greater are the limitations on $\hat{F}_{m,m-1}$ when the remaining $\hat{F}_{pq}$ that enter into the determinant are known. Unfortunately, the formulas simultaneously become more cumbersome.

Let us also consider the amplitudes that compose the determinant of order m = 4. However, let us not take the general case, but study the determinant

$$D_4 = \begin{vmatrix} 1 & & & \\ \hat{F}_{\mathrm{K}-\mathrm{H}} & 1 & & \\ \hat{F}_{\overline{\mathrm{H}}} & \hat{F}_{\overline{\mathrm{K}}} & 1 & \\ \hat{F}_{\mathrm{K}} & \hat{F}_{\mathrm{H}} & \hat{F}_{\mathrm{K}+\mathrm{H}} & 1 \end{vmatrix}.$$

TABLE 13

$\lvert \hat{F}_K \rvert$	$\lvert \hat{F}_H \rvert$									
	0	0.1	0.2	0.3	0.4	0.5	0.6	0.7	0.8	0.9
0	1 −1	0.99 −0.99	0.98 −0.98	0.96 −0.96	0.91 −0.91	0.87 −0.87	0.80 −0.80	0.72 −0.72	0.60 −0.60	0.44 −0.44
0.1	0.99 −0.99	1 −0.98	1 −0.96	0.98 −0.92	0.95 −0.87	0.91 −0.81	0.86 −0.74	0.78 −0.64	0.68 −0.52	0.53 −0.35
0.2	0.98 −0.98	1 −0.96	1 −0.92	0.99 −0.87	0.98 −0.82	0.95 −0.75	0.90 −0.66	0.84 −0.56	0.75 −0.43	0.61 −0.25
0.3	0.96 −0.96	0.98 −0.92	0.99 −0.87	1 −0.82	0.99 −0.75	0.98 −0.68	0.94 −0.58	0.89 −0.47	0.82 −0.34	0.68 −0.14
0.4	0.91 −0.91	0.95 −0.87	0.98 −0.82	0.99 −0.75	1 −0.68	1 −0.60	0.98 −0.50	0.94 −0.38	0.87 −0.23	0.76 −0.04
0.5	0.87 −0.87	0.91 −0.81	0.95 −0.75	0.98 −0.68	1 −0.60	1 −0.50	0.99 −0.39	0.97 −0.27	0.92 −0.12	0.82 0.08
0.6	0.80 −0.80	0.86 −0.74	0.90 −0.66	0.94 −0.58	0.98 −0.50	0.99 −0.39	1 −0.28	1 −0.16	0.96 0	0.89 0.19
0.7	0.72 −0.72	0.78 −0.64	0.84 −0.56	0.89 −0.47	0.94 −0.38	0.97 −0.27	1 −0.16	1 −0.02	0.99 0.13	0.94 0.32
0.8	0.60 −0.60	0.68 −0.52	0.75 −0.43	0.82 −0.34	0.87 −0.23	0.92 −0.12	0.96 0	0.99 0.13	1 0.28	0.98 0.46
0.9	0.44 −0.44	0.53 −0.35	0.61 −0.25	0.68 −0.14	0.76 −0.04	0.82 0.08	0.89 0.19	0.94 0.32	0.98 0.46	1 0.62

Since we have in view a centrosymmetric crystal, $\hat{F}_H = \hat{F}_{\bar{H}}$ and $\hat{F}_K = \hat{F}_{\bar{K}}$.

Equation (85) takes the form

$$\hat{F}_{K+H} = \hat{F}_H \hat{F}_K + \frac{1}{1-\hat{F}^2_{K-H}} (\hat{F}_H - \hat{F}_K \hat{F}_{K-H}) (\hat{F}_K - \hat{F}_H \hat{F}_{K-H}) +$$

$$+ \sqrt{\left[1-\hat{F}^2_K - \frac{1}{1-\hat{F}^2_{K-H}} (\hat{F}_H - \hat{F}_K \hat{F}_{K-K})^2\right]\left[1-\hat{F}^2_H - \frac{1}{1-\hat{F}^2_{K-H}} (\hat{F}_K - \hat{F}_H \hat{F}_{K-H})^2\right]} \cos\varphi \qquad (90)$$

The limiting values of $\hat{F}_{K+H}$ now depend not only on correlations between the absolute values of the amplitudes, but also on the correlation of the two determining signs of the structure products $\hat{F}_H \hat{F}_K \hat{F}_{H+K}$ and $\hat{F}_H \hat{F}_K \hat{F}_{K-H}$ which we shall denote as s_{H+K} and s_{K-H}.

Let us rewrite (90) and multiply both sides of the equation by $\hat{F}_H \hat{F}_K$. For brevity, let us denote $y = \hat{F}_H \hat{F}_K \hat{F}_{H+K}$ and $x = \hat{F}_H \hat{F}_K \hat{F}_{K-H}$. Then

$$y = \hat{F}^2_H \hat{F}^2_K + \frac{1}{1-\hat{F}^2_{K-H}} \left(\hat{F}^2_H - x\right)\left(\hat{F}^2_K - x\right) +$$

$$+ \sqrt{\left[\hat{F}^2_H - \hat{F}^2_H \hat{F}^2_K - \frac{1}{1-\hat{F}^2_{K-H}} \left(\hat{F}^2_H - x\right)^2\right]\left[\hat{F}^2_K - \hat{F}^2_H \hat{F}^2_K - \frac{1}{1-\hat{F}^2_{K-H}} \left(\hat{F}^2_K - x\right)^2\right]} \cos\varphi \qquad (91)$$

In this notation, it is obvious that the limits on the possible values of $\hat{F}_{H+K}$ are determined by the sign s_{K-H}.

Curves of <u>y</u> as a function of <u>x</u> constructed for various values of $|\hat{F}_H|$ and $|\hat{F}_K|$ will prove useful in practical work.[5]

[5]The reliable connections between structure amplitudes studied above may be of substantial advantage in determining the symmetry of the crystal.

Let us give a simple example. A choice must be made between the groups C^4_{2h} (P2/c) and C^5_{2h} ($P2_1/c$).

Let us consider a strong reflection $\hat{F}_{02k0}$ such as would enable one to construct several reliably positive products

$$\hat{F}_{hkl} \hat{F}_{\bar{h}k\bar{l}} \hat{F}_{02k0}.$$

(continued)

18. Graphical Representation of the Connecting Equations for the Simplest D_m

For practical use of the basic connecting equations $D_m > 0$ or, which is the same thing, of (88), graphical methods may be used successfully, though only for the simplest cases (m = 3, m = 4).

Our goal is to determine the signs of structure products for given absolute values $|\hat{F}_{pq}|$.

For m = 3, we obtain from (83) the condition for which the structure product $\hat{F}_{\mathbf{K}-\mathbf{H}}\hat{F}_{\mathbf{H}}\hat{F}_{\mathbf{K}}$ is positive.

$$\hat{F}_{21}\hat{F}_{31} - \sqrt{(1-\hat{F}_{21}^2)(1-\hat{F}_{31})^2} < \hat{F}_{32} < \hat{F}_{21}\hat{F}_{31} + \\ + \sqrt{(1-\hat{F}_{21}^2)(1-\hat{F}_{31}^2)}.$$

This shows that the modulus of $\hat{F}_{32}$ will always be between the limits

$$|\hat{F}_{21}\hat{F}_{31}| \pm \sqrt{(1-\hat{F}_{21}^2)(1-\hat{F}_{31}^2)}.$$

This means that $\hat{F}_{\mathbf{K}-\mathbf{H}}$ must be greater in absolute magnitude than the lower limit of the possible values.

In the positive quadrant of Fig. 15 are constructed the curves

$$z = \left|\, |\hat{F}_{\mathbf{H}}\hat{F}_{\mathbf{K}}| - \sqrt{(1-\hat{F}_{\mathbf{H}}^2)(1-\hat{F}_{\mathbf{K}}^2)}\,\right|. \tag{92}$$

With the aid of this graph it is easy to say whether a structure product is reliably positive. To do this the curve $z = |\hat{F}_{\mathbf{K}-\mathbf{H}}|$ must be chosen on the graph and a point with the coordinates $|\hat{F}_{\mathbf{H}}|$ and $|\hat{F}_{\mathbf{K}}|$ must be marked. If this point lies above the curve, then the structure product is reliably positive.

In the group P2/c the relations $\hat{F}_{hkl} = \hat{F}_{\bar{h}k\bar{l}}$ when $l = 2n$ and $\hat{F}_{hkl} = -\hat{F}_{\bar{h}k\bar{l}}$ when $l \neq 2n$. Whence it is clear that, depending on the sign of $\hat{F}_{02k0}$, amplitudes with either $l = 2n$, or $l \neq 2n$ may enter structure products of the type described.

In the group $C_{2h}^5(P2_1/c)$ the reflections divide up into two further groups ($k + l$ odd and even). Thus, by analyzing the indices of the amplitudes which form strong products $\hat{F}_{hkl}\hat{F}_{\bar{h}k\bar{l}}\hat{F}_{02k0}$, it is always possible to distinguish a simple axis of symmetry from a screw-axis.

The region of the graph above the curve z = 0 gives reliably positive structure products at any value of $\hat{F}_{K-H}$.

We indicated above a simple method of determining positive structure products: all $|\hat{F}_H\hat{F}_K\hat{F}_{K-H}| > \frac{1}{8}$ are reliably positive. It is not hard to see, by considering the figure, that this equation has fewer possibilities than the graph, and thus is less general than the direct consequences of the basic connecting equation.

The connecting equation for m = 4 leads to a cumbersome formula connecting six structure amplitudes. However, it is always possible to impose arbitrary conditions on $\hat{F}_{qp}$. Thus, it is possible to create an unlimited number of inequalities. Take six magnitudes $\hat{F}_{qp}$ that enter the determinant of order m = 4; $\hat{F}_{21}$, $\hat{F}_{31}$, $\hat{F}_{41}$, $\hat{F}_{32}$, $\hat{F}_{42}$ and $\hat{F}_{43}$, and denote them respectively by $\hat{F}_H$, $\hat{F}_K$, $\hat{F}_L$, $\hat{F}_{K-H}$, $\hat{F}_{L-H}$ and $\hat{F}_{L-K}$ and let us assume L = K + H. Then the determinant will give (56) and, of course, also the following from (85)

$$\left(1-\hat{F}^2_{H+K}\right)\left(1-\hat{F}^2_{H-K}\right)+\left(\hat{F}^2_H-\hat{F}^2_K\right)^2 \geqslant$$
$$\geqslant 2\left(\hat{F}^2_H+\hat{F}^2_K\right)\left(1+s_1s_2\left|\hat{F}_{H+K}\hat{F}_{H-K}\right|\right)-4s_1\left|\hat{F}_H\hat{F}_K\hat{F}_{H+K}\right|-$$
$$-4s_2\left|\hat{F}_H\hat{F}_K\hat{F}_{H-K}\right|$$

Inequality (56) under favorable conditions will help us to draw conclusions about the signs s_1 and s_2.

There are four ways of choosing the signs s_1 and s_2. Let us denote the left side of (56) by A, and by a,b,c,d its right side, when choosing respectively the signs − −, + +, − + and + −.

Variant	s_1s_2				
1	− −	$a = 2\left(\hat{F}^2_H+\hat{F}^2_K\right)$	$\left(1+\left\|\hat{F}_{H+K}\hat{F}_{H-K}\right\|\right)$	$+4\left\|\hat{F}_H\hat{F}_K\hat{F}_{H+K}\right\|$	$+4\left\|\hat{F}_H\hat{F}_K\hat{F}_{H-K}\right\|$
2	+ +	$b=$	+	−	−
3	− +	$c=$	−	+	−
4	+ −	$d=$	−	−	+

It is not hard to see that inequality (56) may lead us to six different sign results.

Let us agree always to choose $|\hat{F}_{H+K}| > |\hat{F}_{H-K}|$ – this is always possible, since $\hat{F}_K = \hat{F}_{\bar{K}}$. If this is so, then c > d. Inasmuch as a > b, then the six results just mentioned are as follows (Table 14).

In another combination in addition to the one indicated, sign variants cannot occur in the second column.

For the practical use of inequality (56), it is most convenient to resort to the construction of graphs for the assigned values of $|\hat{F}_{\mathbf{H}}|$ and $|\hat{F}_{\mathbf{K}}|$.

Let us write $x = |\hat{F}_{\mathbf{H}+\mathbf{K}}|$ and $y = |\hat{F}_{\mathbf{H}-\mathbf{K}}|$ and write (56) in the form

$$(1-x^2)(1-y^2) - 2(\hat{F}^2_{\mathbf{H}} + \hat{F}^2_{\mathbf{K}})(1 + s_1 s_2 xy) + \\ + 4|\hat{F}_{\mathbf{H}}\hat{F}_{\mathbf{K}}|(s_1 x + s_2 y) + (\hat{F}^2_{\mathbf{H}} - \hat{F}^2_{\mathbf{K}})^2 \geqslant 0.$$

Denoting the left side of the inequality by z, let us reduce it to the form of a polynomial of the second degree in y. Evidently,

$$z \equiv -y^2(1-x^2) + [4|\hat{F}_{\mathbf{H}}\hat{F}_{\mathbf{K}}|s_2 - 2(\hat{F}^2_{\mathbf{H}} + \hat{F}^2_{\mathbf{K}})s_1 s_2 x]\, y + \\ + [1 - x^2 - 2(\hat{F}^2_{\mathbf{H}} + \hat{F}^2_{\mathbf{K}}) + 4|\hat{F}_{\mathbf{H}}\hat{F}_{\mathbf{K}}|s\ x + (\hat{F}^2_{\mathbf{H}} - \hat{F}^2_{\mathbf{K}})^2]. \qquad (93a)$$

Let us find the roots of the equation $z = 0$. Let them be equal to $y_1 = f_1(x)$ and $y_2 = f_2(x)$. Then the inequality may be written in the form

$$-(y - f_1(x))(y - f_2(x)) \geqslant 0 \qquad (93b)$$

and will acquire the following geometric meaning: On the graph of x, y, where $x = |\hat{F}_{\mathbf{H}+\mathbf{K}}|$ and $y = |\hat{F}_{\mathbf{H}-\mathbf{K}}|$, the curves of $f_1(x)$ and $f_2(x)$, constructed for the given $|\hat{F}_{\mathbf{H}}|$ and $|\hat{F}_{\mathbf{K}}|$, limit the field of the quantities $|\hat{F}_{\mathbf{H}+\mathbf{K}}|$, $|\hat{F}_{\mathbf{H}-\mathbf{K}}|$, for which fulfillment of the inequaltiy is possible [it follows from (93b) that the points y must lie between the curves f_1 and f_2].

Let us find the form of f_1 and f_2, i.e., solve the quadratic equation $z = 0$. We obtain

$$f_1(x) = -1 + (\hat{F}^2_{\mathbf{H}} + 2|\hat{F}_{\mathbf{H}}\hat{F}_{\mathbf{K}}|s_2 + \hat{F}^2_{\mathbf{K}})\frac{1 - s_1 s_2 x}{1 - x^2};$$

$$f_2(x) = 1 - (\hat{F}^2_{\mathbf{H}} - 2|\hat{F}_{\mathbf{H}}\hat{F}_{\mathbf{K}}|s_2 + \hat{F}^2_{\mathbf{K}})\frac{1 + s_1 s_2 x}{1 - x^2}.$$

Introducing the notations $|\hat{F}_{\mathbf{H}}| + |\hat{F}_{\mathbf{K}}| = E$ and $||\hat{F}_{\mathbf{H}}| - |\hat{F}_{\mathbf{K}}|| = G$, we obtain for the four sign variants the following equations of hyperbolas:

$$s_1 = -1; \quad s_2 = -1$$

$$f_1(x) = -1 + \frac{G^2}{1+x}; \quad (1+y)(1+x) = G^2;$$

$$f_2(x) = 1 - \frac{E^2}{1-x}; \quad (1-y)(1-x) = E^2;$$

$$s_1 = +1; \quad s_2 = +1$$

$$f_1(x) = -1 + \frac{E^2}{1+x}; \quad (1+y)(1+x) = E^2,$$

$$f_2(x) = 1 - \frac{G^2}{1-x}, \quad (1-y)(1-x) = G^2,$$

$$s_1 = -1; \quad s_2 = +1$$

$$f_1(x) = -1 + \frac{E^2}{1-x}; \quad (1+y)(1-x) = E^2;$$

$$f_2(x) = 1 - \frac{G^2}{1-x}; \quad (1-y)(1+x) = G^2;$$

$$s_1 = +1; \quad s_2 = -1$$

$$f_1(x) = -1 + \frac{G^2}{1-x}; \quad (1+y)(1-x) = G^2;$$

$$f_2(x) = 1 - \frac{E^2}{1+x}; \quad (1-y)(1+x) = E^2,$$

Thus, the variations in the signs s_1, s_2 are solved on condition that the values $|\hat{F}_{H-K}|$ lie on the graph $|\hat{F}_{H-K}| = f(|\hat{F}_{H+K}|)$ between the curves indicated.

For the practical use of inequality (56), or (85), which, in general, is the same thing, one may construct the regions of sign results 0, I,..., V for all pairs of values of $|\hat{F}_H|$ and $|\hat{F}_K|$ through 0.2 (this is quite sufficient, if one has in view the accuracy of the measurements) in the field of values of $\underline{x}$, $\underline{y}$ that are of interest to us. The field $\underline{x}$, $\underline{y}$ is a triangle bounded by the lines $y = 0$, $x = 1$, and $y = x$ (we assumed $x > y$). Figure 16 shows four families of hyperbolas passing through this triangle.

Knowing G and E for each pair of values $|\hat{F}_H|$ and $|\hat{F}_K|$, we superimpose all the hyperbolas that fall within the field of one triangle. Aware where the hyperbolas pass (higher or lower than the graph) that have gone beyond the limits of the field, we may sketch in the region of variation of the choice of the signs s_1 and s_2. These regions overlap in a number of cases. Such overlapping leads either to one of the sign variants 0, I,... V, or to an impossible region.

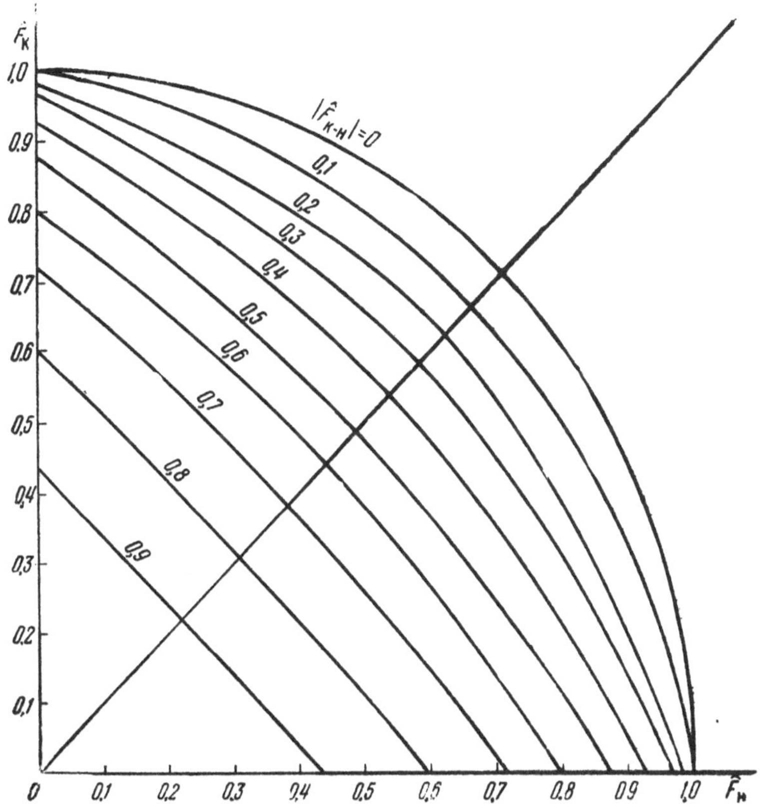

Fig. 15. Curves for the determination of the positive sign of $\hat{F}_{\mathbf{H}}\hat{F}_{\mathbf{K}}\hat{F}_{\mathbf{K-H}}$.

TABLE 14

Magnitude	Variations which are solved	Result of applying inequality	Number of resulting signs
$A > a, b, c, d$	1—4	Arbitrary result	0
$a, c, d > A > b$	2	$s_1 = s_2 = +1$	I
$a, b, c > A > d$	4	$s_1 = +1,\ s_2 = -1$	II
$a, c > A > b, d$	2,4	$s_1 = +1$	III
$a, b > A > c, d$	3,4	$s_1 = -s_2$	IV
$a > A > b, c, d$	2—4	Any result, except $s_1 = s_2 = -1$	V

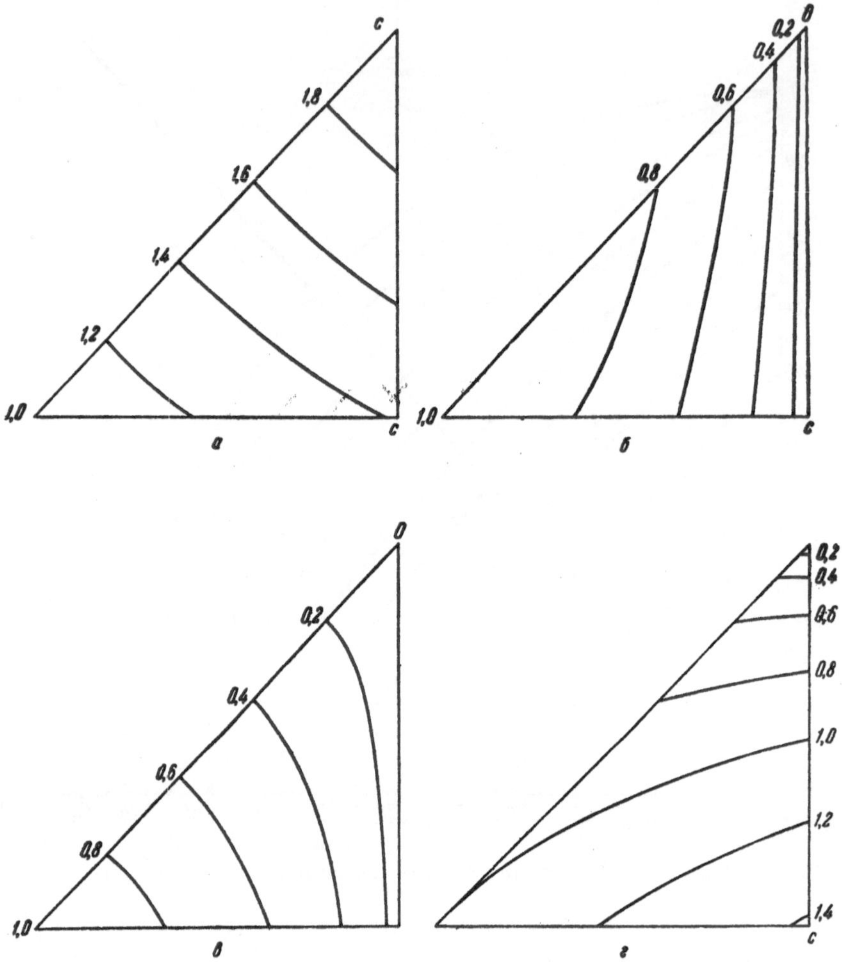

Fig. 16. Pertaining to the graphic method of studying the inequality $D_4 > 0$. a) $(1 + y)(1 + x) = c^2$; b) $(1 + y)(1 - x) = c^2$; c) $(1 - y)(1 - x) = c^2$; d) $(1 - y)(1 + x) = c^2$.

Figure 17 shows the detailed construction of the region for $|\hat{F}_H| = 0.7$ and $|\hat{F}_K| = 0.1$. Figure 18 shows the working graphs for all possible values of the structure amplitudes. It must be remembered that if both limiting hyperbolas lie beyond the limits of the triangle, both "below" or both "above" it, then the corresponding sign variant is forbidden. On the contrary, if the limiting hyperbolas embrace the triangle, then the corresponding sign variant of s_1 and s_2 is allowed in the whole field.

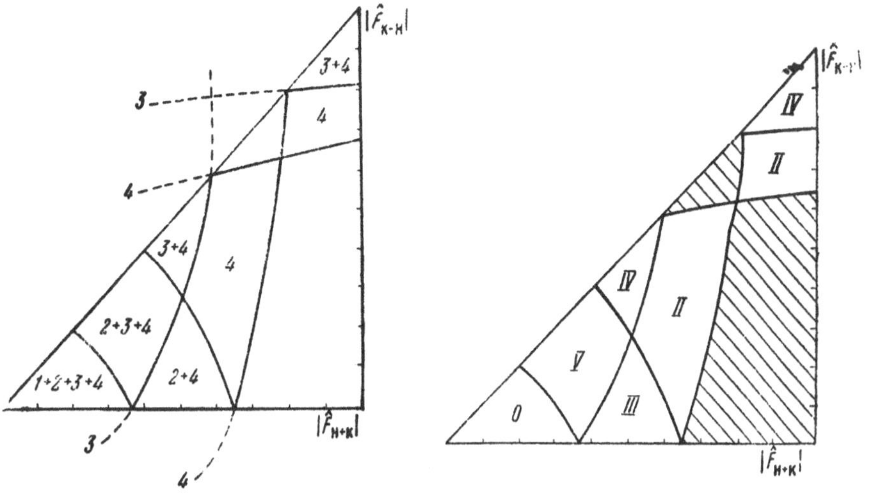

Fig. 17. Construction of the graph for the determination of signs $\hat{F}_{\mathbf{K}+\mathbf{H}}$ and $\hat{F}_{\mathbf{K}-\mathbf{H}}$ and $|\hat{F}_{\mathbf{H}}| = 0.7$ and $|\hat{F}_{\mathbf{K}}| = 0.1$.

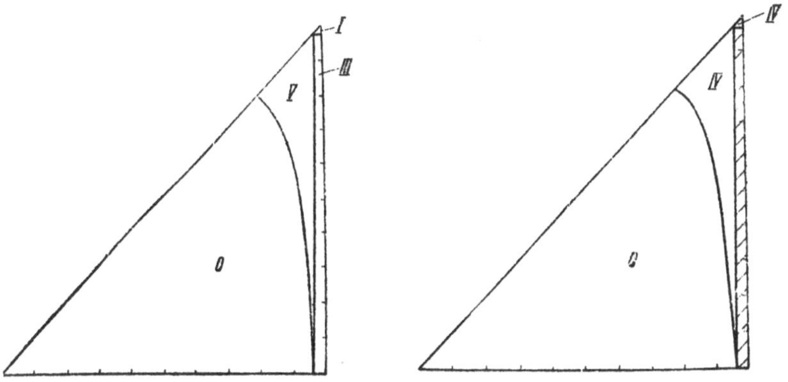

1. $|\hat{F}_{\mathbf{H}}| = 0{,}1$, $|\hat{F}_{\mathbf{K}}| = 0{,}1$;

2. $|\hat{F}_{\mathbf{H}}| = 0{,}2$, $|\hat{F}_{\mathbf{K}}| = 0$;

Fig. 18. (1-23). Working graphs for the determination of signs of $\hat{F}_{\mathbf{K}+\mathbf{H}}$ and $\hat{F}_{\mathbf{K}-\mathbf{H}}$ for given $|\hat{F}_{\mathbf{H}}|$ and $|\hat{F}_{\mathbf{K}}|$.

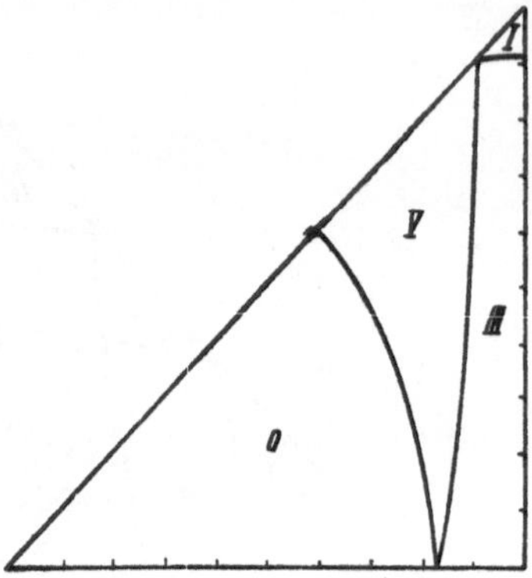

3. $|\hat{F}_{\mathbf{H}}| = 0.2$, $|\hat{F}_{\mathbf{K}}| = 0.2$;

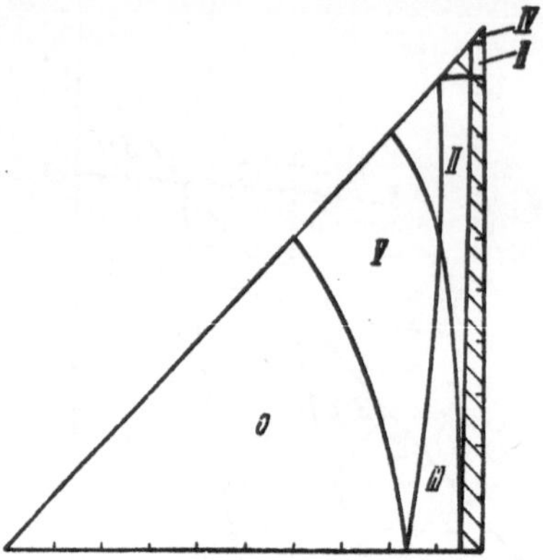

4. $|\hat{F}_{\mathbf{H}}| = 0.3$, $|\hat{F}_{\mathbf{K}}| = 0.1$;

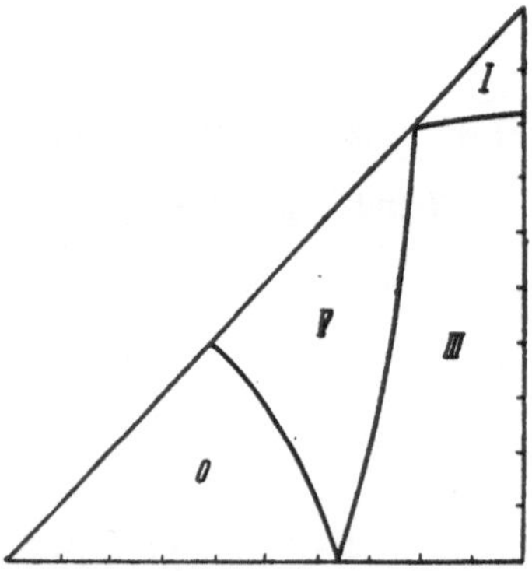

5. $|\hat{F}_{\mathbf{H}}| = 0.3$, $|\hat{F}_{\mathbf{K}}| = 0.3$;

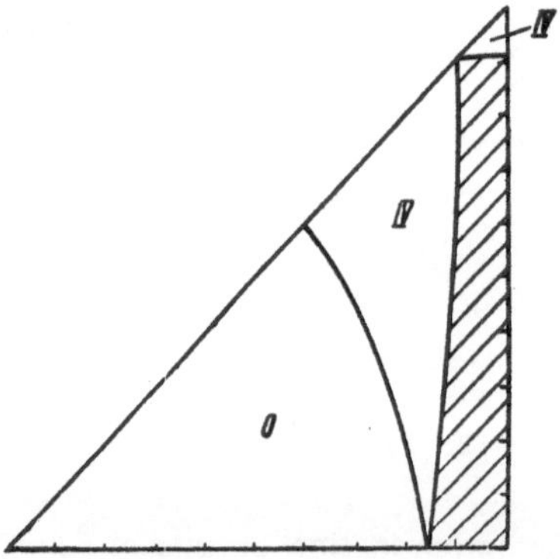

6. $|\hat{F}_{\mathbf{H}}| = 0.4$, $|\hat{F}_{\mathbf{K}}| = 0$;

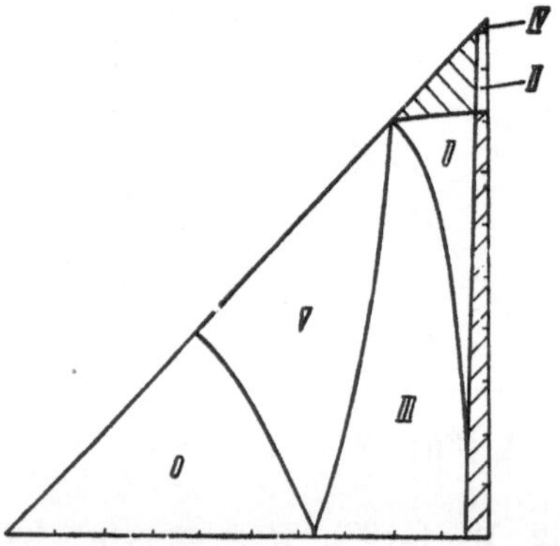

7. $|\hat{F}_{\mathbf{H}}| = 0.4$, $|\hat{F}_{\mathbf{K}}| = 0.2$;

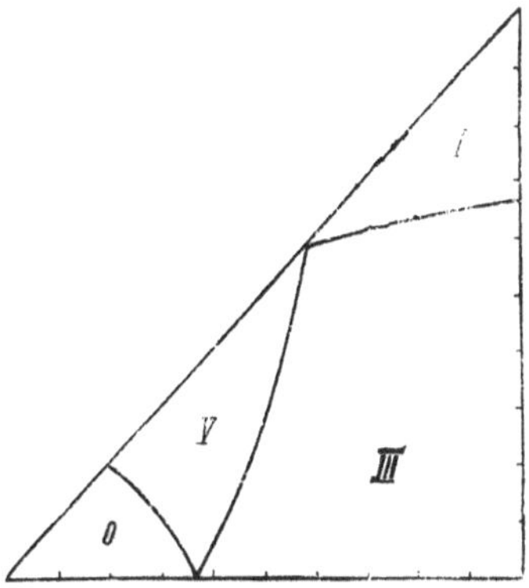

8. $|\hat{F}_{\mathbf{H}}| = 0.4$, $|\hat{F}_{\mathbf{K}}| = 0.4$;

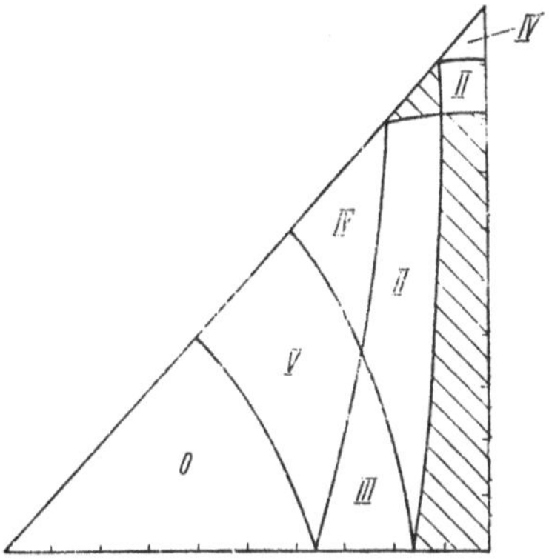

9. $|\hat{F}_{\mathbf{H}}| = 0.5$, $|\hat{F}_{\mathbf{K}}| = 0.1$;

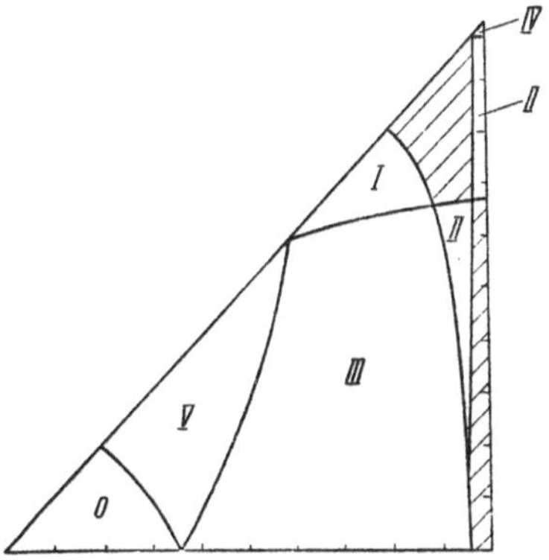

10. $|\hat{F}_{\mathbf{H}}| = 0.5$, $|\hat{F}_{\mathbf{K}}| = 0.3$;

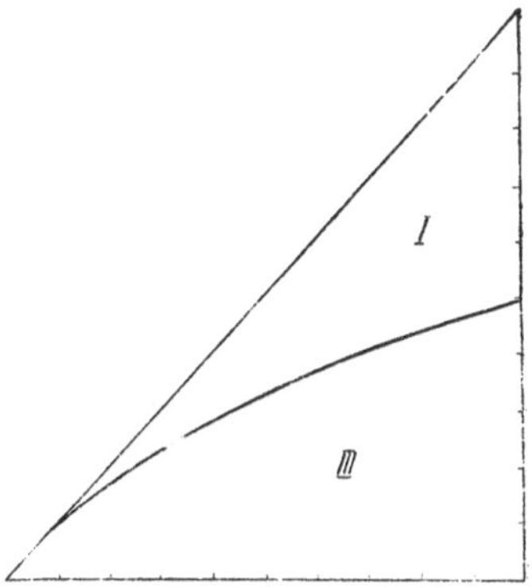

11. $|\hat{F}_{\mathbf{H}}| = 0.5$, $|\hat{F}_{\mathbf{K}}| = 0.5$;

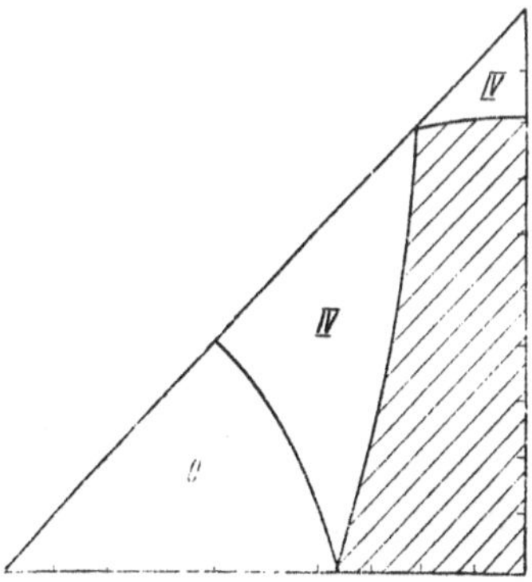

12. $|\hat{F}_{\mathbf{H}}| = 0.6$, $|\hat{F}_{\mathbf{K}}| = 0$;

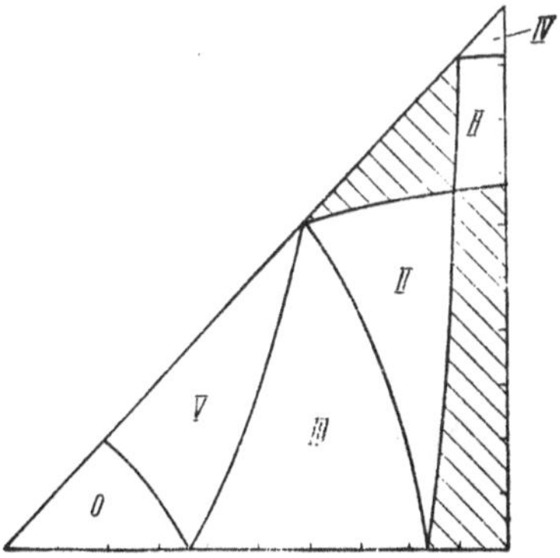

13. $|\hat{F}_{\mathbf{H}}| = 0.6$, $|\hat{F}_{\mathbf{K}}| = 0.2$,

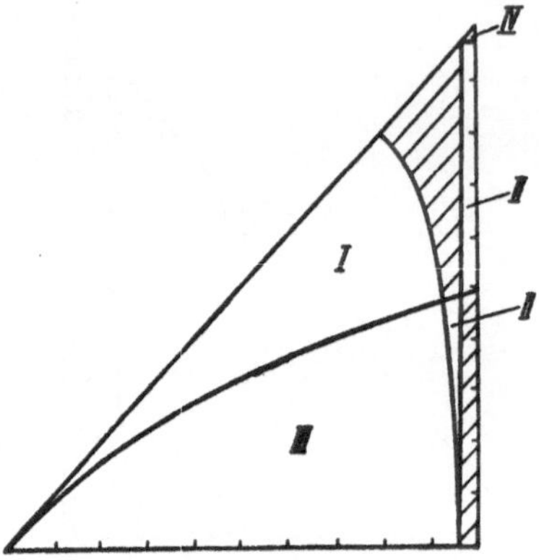

14. $|\hat{F}_{\mathbf{H}}| = 0{,}6$, $|\hat{F}_{\mathbf{K}}| = 0.4$;

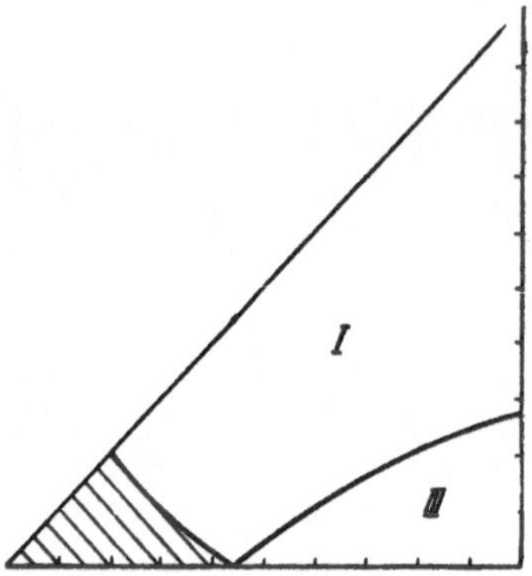

15. $|\hat{F}_{\mathbf{H}}| = 0{,}6$, $|\hat{F}_{\mathbf{K}}| = 0{,}6$;

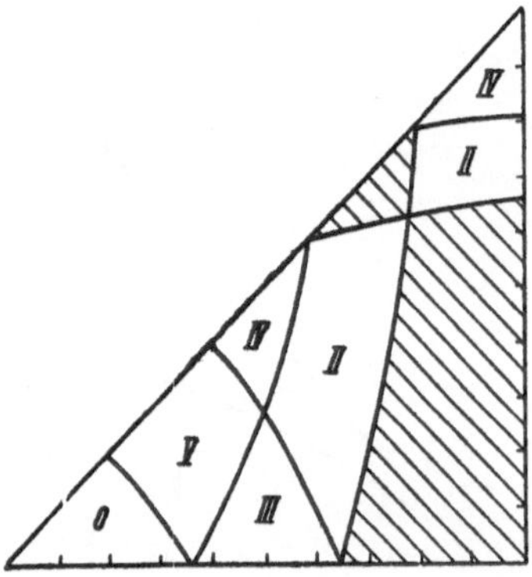

16. $|\hat{F}_{\mathbf{H}}| = 0{,}7$, $|\hat{F}_{\mathbf{K}}| = 0.1$;

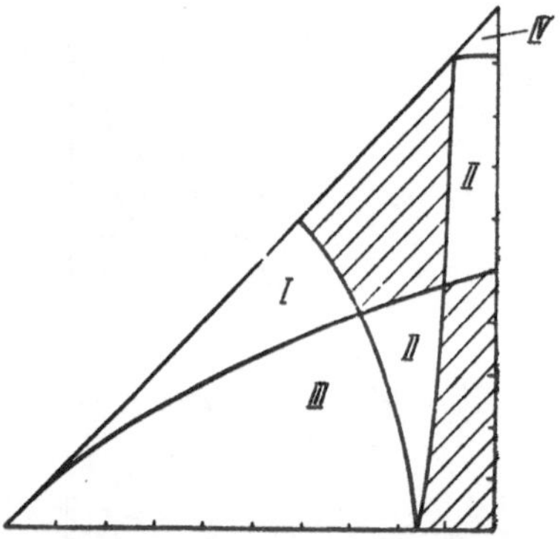

17. $|\hat{F}_{\mathbf{H}}| = 0.7$, $|\hat{F}_{\mathbf{K}}| = 0.3$;

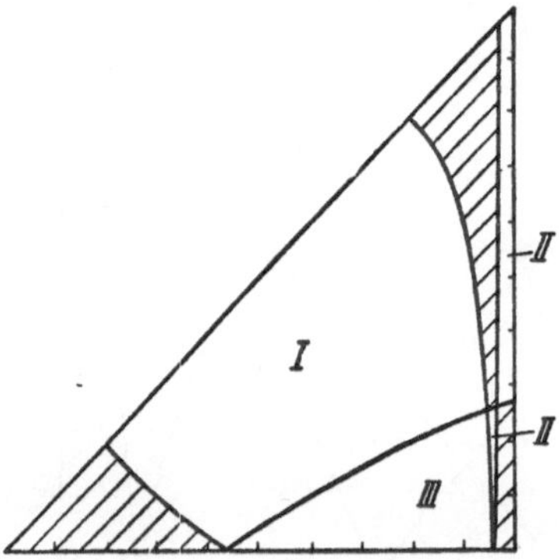

18. $|\hat{F}_{\mathbf{H}}| = 0.7$, $|\hat{F}_{\mathbf{K}}| = 0.5$;

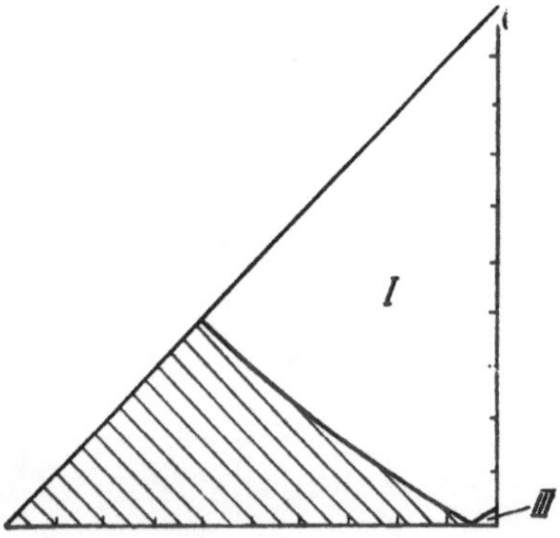

19. $|\hat{F}_{\mathbf{H}}| = 0.7$, $|\hat{F}_{\mathbf{K}}| = 0.7$;

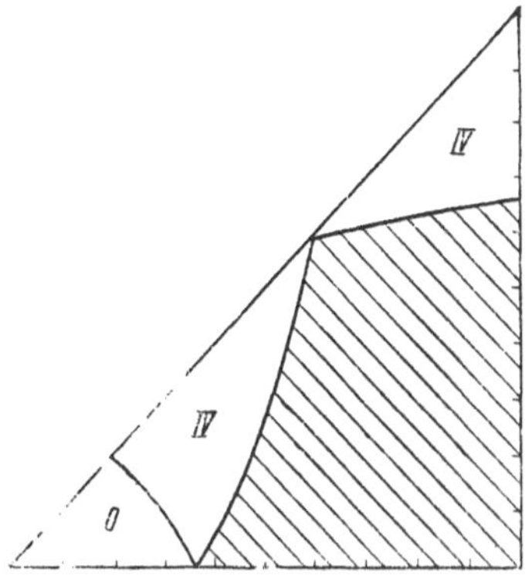

20. $|\hat{F}_{\mathbf{H}}| = 0.8$, $|\hat{F}_{\mathbf{K}}| = 0$;

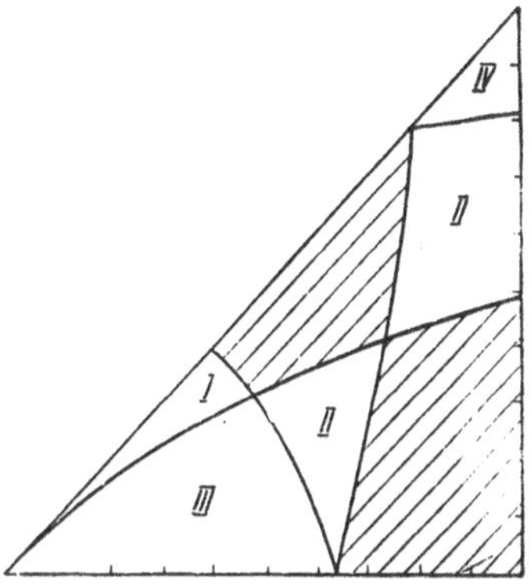

21. $|\hat{F}_{\mathbf{H}}| = 0.8$, $|\hat{F}_{\mathbf{K}}| = 0.2$;

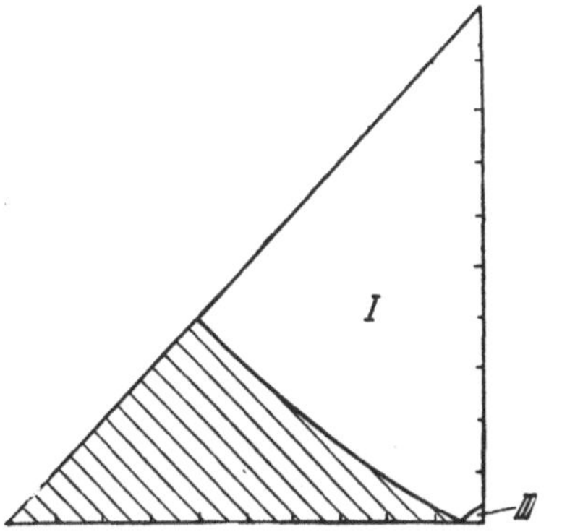

22. $|\hat{F}_{\mathbf{H}}| = 0.8$, $|\hat{F}_{\mathbf{K}}| = 0.6$;

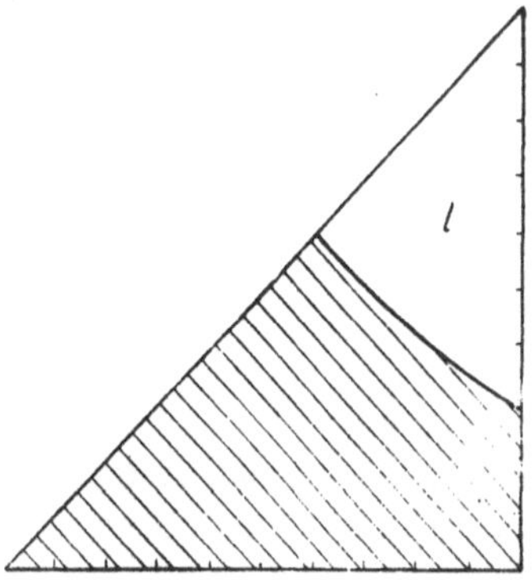

23. $|\hat{F}_{\mathbf{H}}| = 0.8$; $|\hat{F}_{\mathbf{K}}| = 0.8$.

19. The Complete Theory of the Relationships Between Structure Amplitudes [40]

From the content of this chapter, the character of the connections between structure amplitudes is obvious. If one, two, or several structure amplitudes are given, then the magnitudes of the other amplitudes, generally speaking, are not arbitrary. In its clearest form, this dependence is given by the basic connecting equation (48) and its consequence (86):

$$\hat{F}_{m,\, m-1} = A + B \cos \varphi,$$

where A and B are expressed in terms of the remaining elements of the determinant with the elements $\hat{F}_{pq}$ and φ is a parameter which may take any value between 0 and 2π.

Thus, there is no unique connection between all the $\hat{F}_{pq}$ and a single element $\hat{F}_{m,m-1}$. For any given $\hat{F}_{pq}$, except one, this element has values in the interval A ± B. To each value of φ there corresponds some value of $F_{m,m-1}$.

Let us face the problem concerning the probability of the structure where amplitudes are $\hat{F}_{pq}$.

If all the amplitudes, except $\hat{F}_{m,m-1}$, are assigned, then the probability of such a structure depends only on one parameter φ. Let us denote the number of structures with parameters between φ and $\varphi + d\varphi$ by $P(\varphi)d\varphi$. Since φ and $\hat{F}_{m,m-1}$ are uniquely connected, this probability is equal to $\eta\,(\hat{F})\,d\hat{F}$, i.e., to the number of times that $\hat{F}_{m,m-1}$ lies in the interval between $\hat{F}$ to $\hat{F} + d\hat{F}$:

$$P(\varphi)\,d\varphi = \eta\,(\hat{F})\,d\hat{F}, \tag{94}$$

consequently,

$$P(\varphi) = \eta\,(\hat{F})\frac{1}{\frac{d\varphi}{d\hat{F}}} = \text{const}\,\sqrt{B^2 - (\hat{F}_{m,\,m-1} - A)^2}\,\eta\,(\hat{F}). \tag{95}$$

With respect to η, its exact computation is probably difficult. We shall accept for η a Gaussian distribution centered at A (the mean value of $\hat{F}_{m,m-1} = A$) and having a dispersion indistinguishable from that for the general case (when there are no fixed amplitudes):

$$\eta \sim \exp - \frac{(\hat{F}_{m,\,m-1} - A)^2}{2\sum n_j^2}. \tag{96}$$

Thus, we come to the following basic formula:

$$P(\hat{F}_{m,\,m-1}) = \text{const}\,\sqrt{B^2 - (\hat{F}_{m,\,m-1} - A)^2}\,\exp - \frac{(\hat{F}_{m,\,m-1} - A)^2}{2\sum n_j^2}. \tag{97}$$

This is the formula for the probable value of $\hat{F}_{m,m-1}$ for given values of the remaining elements $\hat{F}_{pq}$ of the connecting determinant of the $\underline{m}$th order.

From the theory just stated, it is evident that one can speak of the probable value of this or that structure amplitude only when indicating whether the other amplitudes are assumed to be known and, in addition, what kind and how many.

Unfortunately, (97) is not proved strictly [it is necessary to prove, first, the independence of P_φ and η, and secondly, (96)].

Nevertheless, it is doubtful whether a strict theory, if it be constructed, will produce substantial corrections in (97). In any case, the results from (97) stated in the following paragraph coincide well with experiment and give correct results for extreme cases which have been strictly proven.

20. The Probability of a Positive Sign of the Structure Product

We find from (97) the probability of a positive sign of the structure product $\hat{F}_{m,m-1}\,\hat{F}_{m,1}\,\hat{F}_{m-1,1}$; forming

$$W'_+ = \frac{P_+}{P_+ + P_-} = \frac{1}{1+\varepsilon}, \tag{98}$$

where $\epsilon = \dfrac{P_-}{P_+}$, we obtain

$$\varepsilon = \sqrt{\frac{B^2 - (-\hat{F}_{m,\,m-1} - A)^2}{B^2 - (\hat{F}_{m,\,m-1} - A)^2}}\,\exp - \frac{2\hat{F}_{m,\,m-1}A}{\sum n_j^2}. \tag{99}$$

For the simplest case of a determinant of order m = 3 we have

$$\varepsilon = \sqrt{\frac{1 - \hat{F}_{\mathrm{H}}^2 - \hat{F}_{\mathrm{K}}^2 - \hat{F}_{\mathrm{K-H}}^2 - 2\,|\hat{F}_{\mathrm{H}}\hat{F}_{\mathrm{K}}\hat{F}_{\mathrm{K-H}}|}{1 - \hat{F}_{\mathrm{H}}^2 - \hat{F}_{\mathrm{K}}^2 - \hat{F}_{\mathrm{K-H}}^2 + 2\,|\hat{F}_{\mathrm{H}}\hat{F}_{\mathrm{K}}\hat{F}_{\mathrm{K-H}}|}}\,\exp - \frac{2\,|\hat{F}_{\mathrm{H}}\hat{F}_{\mathrm{K}}\hat{F}_{\mathrm{K-H}}|}{\sum n_j^2}. \tag{100}$$

For small structure products, i.e., for those (see Chapter III), which are smaller than $\sim 10\left(\sum n_j^2\right)^{3/2}$, the last formula gives

$$\varepsilon \approx \exp - \frac{2\,|\hat{F}_{\mathrm{H}}\hat{F}_{\mathrm{K}}\hat{F}_{\mathrm{K-H}}|}{\sum n_j^2}. \tag{101}$$

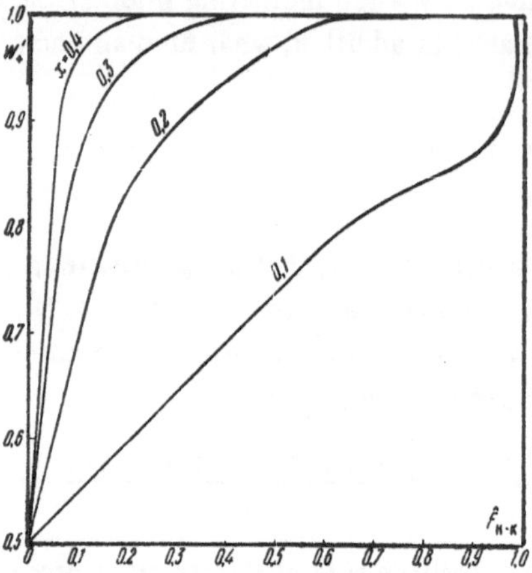

Fig. 19. Curves of the probability W_+ for $\Sigma n_j^2 = 0{,}1$, $N_{eff} = 100$, $x = \hat{F}_{\mathbf{H}} = \hat{F}_{\mathbf{K}}$.

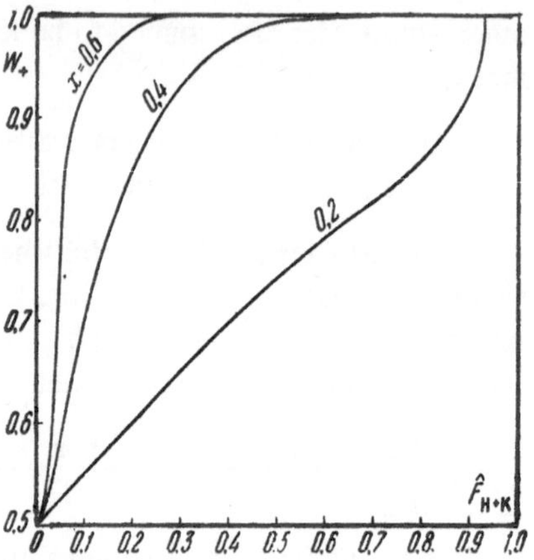

Fig. 20. Curves of the probability W_+ for $\Sigma n_j^2 = 0{,}2$, $N_{eff} = 25$, $x = \hat{F}_{\mathbf{H}} = \hat{F}_{\mathbf{K}}$.

This formula differs (not substantially, it is true) from the formula for W_+ derived in chapter III for small structure products; which is natural, since W_+' has a somewhat different meaning from that of W_+. The first probability calculated from (97) is the probability of the sign of $\hat{F}_{\mathbf{K}-\mathbf{H}}\hat{F}_{\mathbf{K}}\hat{F}_{\mathbf{H}}$ for a fixed value of $\hat{F}_{\mathbf{H}}\hat{F}_{\mathbf{K}}$. The probability W_+ studied in chapter III was calculated without assuming any fixed amplitudes.

A formula for W_+' suitable for small structure products has been derived by Wolfson.

A working graph of the useful function

$$W'_+ = \frac{1}{1+\varepsilon},$$

where ϵ is given by (101), is shown in Figures 19 and 20 for a series of values $\hat{F}_{\mathbf{H}}\hat{F}_{\mathbf{K}}$ and for two values of $\sum n_j^2$.

For determinants of higher orders the formulas become cumbersome and difficult to analyze.

It is not hard to construct a probability formula for the positive sign of the structure product $\hat{F}_H\hat{F}_K\hat{F}_{K+H}$ for given absolute values of $\hat{F}_H$, $\hat{F}_K$, $\hat{F}_{K+H}$ and $\hat{F}_{K-H}$, and for a given sign $\underline{s}$ of the structure product $\hat{F}_H\hat{F}_K\hat{F}_{K-H}$. In this case one must use the special connecting determinant for m = 4 (see page 186), i.e., (90), and substitute the values of A and B from (90) into (99).

21. Procedures for Direct Structure Analyses

The theory of the connections between structure amplitudes will have a full and broad application in structure analysis when there are rapid computing machines at the researcher's disposal. Therefore, it is quite probable that future experience will bring substantial corrections to the methods of structure analysis which we present.

Before starting the analysis the researcher should compute the mean value of the structure amplitude — the quantity $\sigma = (\sum n_j^2)^{1/2}$.

The value of σ determines in the first place the possibilities of the structure analysis and predetermines the scheme which must be followed in order to carry it out. The presence or absence in the structure of a heavy atom is of no importance in itself. The presence of a heavy atom allows the successful application of the method of the convolution of the electron density. As to the method of sign determination, its possibilities, we repeat, are determined only by the value of σ.

Scheme I. $\sigma > 0.2$, i.e., $N_{eff} < 25$.

1. We distribute the measured amplitudes $|\hat{F}|$ in a series according to decreasing magnitude. We find the values of $\hat{F}$ by using some approximate $\underline{f}$ curve.

2. With the aid of the table of the Gaussian function (page 86) we separate out the $\hat{F} > 1/2$ (the strong $|\hat{F}|$).

3. We extend the table of structure amplitudes greater than $1/2$ by including in it the symmetrically related ones.

4. We form structure products by combining three strong amplitudes. These are reliably positive.

5. We choose, in accordance with the rules discussed on page 180, the signs of the three amplitudes.

6. From (4) we find the signs and also the correlations between the signs of the strong amplitudes.

7. We form structure products from all the pairs of strong amplitudes.

8. We write out the structure products into which enter one and the same weak amplitude. Considering the structure products to be positive, we find the signs of the weak $\hat{F}$'s which enter into them. If there are discrepant results we postpone the sign determinations of such amplitudes.

9. If it is desirable to determine a greater number of signs, we form structure products between one strong and two weak amplitudes and repeat (8).

Scheme II. $0.2 > \sigma > 0.15$.

1. We determine the unitary structure amplitudes with the aid of the procedure explained in Chapter III, and we distribute the $\hat{F}$'s in a series of descending magnitudes, extending the table of $\hat{F}$'s by using the symmetry of the crystal.

2. We form the reliably positive structure products of $\hat{F} > 1/2$.

3. We form an additional number of reliably positive structure products with the aid of Table 13 on page 187 or the curves in Fig. 15.

4. If the number of reliably positive structure products is sufficient to determine the signs of ten to thirty amplitudes and to obtain a certain number of correlations between the signs of other amplitudes, then the further course of the analysis coincides with the first scheme. Otherwise, one must proceed according to schemeIII.

Scheme III. $0.15 > \sigma > 0.10$.

1-3. The same as for scheme II.

4. We seek a connection between the amplitudes by studying successively the determinants D_m for $m > 3$. For this we sort out successively all the triplets of amplitudes $\hat{F}_{\mathbf{H}}$, $\hat{F}_{\mathbf{K}}$, $\hat{F}_{\mathbf{L}}$ (starting, of course, with the strongest) and on this basis we construct D_4. Then we sort out all the quadruplets $\hat{F}_{\mathbf{H}}$, $\hat{F}_{\mathbf{K}}$, $\hat{F}_{\mathbf{L}}$, and $\hat{F}_{\mathbf{M}}$, and on their basis we construct D_5. From the condition of the positivity of D_m, correlations will be found between the signs of structure products, and it is possible to find among them reliably positive ones. (See pages 171 and 178). Computation by the formulas given on page 186 is also possible. The procedure is cumbersome and easily realized in practice only by the aid of rapid computing machines.

Further Remarks on Scheme II

Karle and Hauptman [19], and also Rumanova [44] have offered schemes for sign determination, although they did not use the reliable connections between structure amplitudes. However, these schemes can succeed only when there is among the structure products a large number of reliably positive ones. In other words, these schemes are nothing but a camouflaged application of our scheme I.

Methods of direct sign determination as yet have had negligible application, which is explained first of all by a poor understanding of the essence of the subject.

The most successful systematic use of direct methods of sign determination for finding complex structures is being carried out by the school of Academician N. V. Belov.

CHAPTER V

ANALYSIS OF THE CONVOLUTION OF THE ELECTRON DENSITY

1. The Convolution as a Sum of Interatomic Functions

As has been shown in Chapter II, the self-convolution of the electron density $S(\rho_0^i, \rho_0)$ may be computed directly from an x-ray experiment. This function, periodically extended to infinity, is called a Patterson function, and is computed experimentally from the formula

$$P(r) = \frac{1}{V} \sum F^2 \exp - 2\pi i \mathbf{H}\mathbf{r},$$

where $P(r) = S\left[S(\rho_0, \rho_0), \frac{G}{z}\right]$, and G is the lattice function (see Chapter I).

Let us represent the density in the cell as a sum of atomic functions

$$\rho_0 = \sum_{j=1}^{N} \rho_j (\mathbf{r} - \mathbf{r}_j).$$

Then

$$S(\rho_0^i, \rho_0) = \int \rho_0 (\mathbf{r}') \rho_0 (\mathbf{r}' - \mathbf{r}) \, dv'$$

($\mathbf{r}$ can also be positive, since the self-convolution is centrosymmetric).

Substituting the value of ρ, we obtain

$$S(\rho_0^i, \rho_0) = \sum_j \sum_{j'} \int \rho_j (\mathbf{r}' - \mathbf{r}_j) \rho_{j'} (\mathbf{r}' - \mathbf{r} - \mathbf{r}_{j'}) \, dv'.$$

The integral

$$\chi_{jj'} (\mathbf{r}) = \int \rho_j (\mathbf{r}' - \mathbf{r}_j) \rho_{j'} (\mathbf{r}' - \mathbf{r} - \mathbf{r}_{j'}) \, dv\cdot$$

may be called an interatomic function of the atoms j and j'.

In the case of a crystal with a center of inversion

$$\rho_{j'}(\mathbf{r}' - \mathbf{r} - \mathbf{r}_{j'}) = \rho_{j'}[\mathbf{r} - (\mathbf{r}' - \mathbf{r}_{j'})],$$

and consequently,

$$\chi_{jj'} = S(\rho_{j'}, \rho_j) = \int \rho_j(\mathbf{r}' - \mathbf{r}_j)\,\rho_{j'}[\mathbf{r} - (\mathbf{r}' - \mathbf{r}_{j'})]\,dv'.$$

Since ρ_j functions are bell shaped, it is evident from (1) that $\chi_{jj'}$ must have a maximum at $\mathbf{r} = \mathbf{r}_j - \mathbf{r}_{j'}$, i.e., when $\mathbf{r}$ is equal to an interatomic vector. This takes place under the assumption that the maximum of $\chi_{jj'}(\mathbf{r})$ lies at the same value of $\mathbf{r}$ as the maximum of the function under the integral ($\mathbf{r}'_j = \mathbf{r}_j = \mathbf{r} - \mathbf{r}_{j'}$, whence $\mathbf{r} = \mathbf{r}_j - \mathbf{r}_{j'}$).

Since the Fourier transform of $\rho_j(\mathbf{r} - \mathbf{r}_j)$ and of $\rho_{j'}(\mathbf{r} - \mathbf{r}_{j'})$ are the atomic factors f_j and $f_{j'}$, we may use (I, 35), which gives

$$\chi_{jj'}(\mathbf{R}) = \Phi(f_j f_{j'}). \tag{1}$$

This proves the following important theorem: An interatomic function is the Fourier transform of a product of atomic factors. It can be said that $f_j f_{j'}$ plays the role of an atomic factor for the interatomic function $\chi_{jj'}$.

The convolution of the electron density is evidentally a sum of interatomic functions

$$S(\rho_0, \rho_0) = \sum_j \sum_{j'} \chi_{jj'}(\mathbf{R} - \mathbf{r}_{jj'}). \tag{2}$$

It is expedient to isolate the terms with $j = j'$

$$S(\rho_0, \rho_0) = \sum_j \chi_{jj}(\mathbf{R}) + \sum_{j \neq j'}\sum \chi_{jj'}(\mathbf{R} - \mathbf{r}_{jj'}). \tag{3}$$

χ_{jj} is the Fourier transform of the square of an atomic factor

$$\chi_{jj}(\mathbf{R}) = \Phi(f_j^2). \tag{4}$$

The single sum in the expression for $S(\rho_0, \rho_0)$ is the so-called zero maximum of the Patterson function. The remaining maxima of the Patterson function would lie exactly at coordinates coinciding with the ends of interatomic vectors in the structure, if there were no strong mutual superposition.

2. The Form of the Interatomic Function

Knowing the approximate shape of the atomic factor, it is not difficult to compute the form of the interatomic function.

If the atomic factor is approximated by the expression $f_j = k_j e^{-\alpha s}$, where $s = 2\pi\,|H|$, then

$$f_j f_{j'} = k_j k_{j'} e^{-(\alpha_j + \alpha_{j'})s}.$$

In this case (see Chapter I)

$$\chi_{jj'}(R) = \frac{k_j k_{j'}(\alpha_j + \alpha_{j'})}{\pi^2\left((\alpha_j + \alpha_{j'})^2 + R^2\right)^2}. \tag{5}$$

If the atomic factor is represented in the form

$$f = Z_j e^{-\beta_j s^2},$$

then

$$f_j f_{j'} = Z_j Z_{j'} e^{-(\beta_j + \beta_{j'})s^2}.$$

In which case

$$\chi_{jj'}(R) = Z_j Z_{j'} \frac{e^{-\frac{r^2}{4(\beta_j + \beta_{j'})}}}{8\left[\pi(\beta_j + \beta_{j'})\right]^{3/2}} \tag{6}$$

We see that the height of a maximum of the interatomic function is proportional to the product of the atomic numbers of the corresponding atoms.

Let us note that a comparison of the peak heights of two interatomic functions will give the ratio $Z_j Z_{j'} / Z_{j_1} Z_{j_1'}$ only when the corresponding atomic curves are alike, i.e., if the coefficients α_j (or β_j) are identical for the atoms under study.

The second important circumstance which should be stressed is the diffuseness of the interatomic function in comparison with the atomic function. It may be said of the second of the approximations under study, that the half width of the Gaussian curve of an interatomic function is equal to the square root of the sum of the squares of the half widths of the Gaussian functions representing the densities of the corresponding atoms.

This leads to strong overlapping of interatomic functions in the Patterson series P (r), which complicates its interpretation. For the same reason, the study of the form of interatomic functions, which would be valuable in judging the shape of the atomic factors, is hardly possible.

Some information about the root mean square of the atomic factors may be obtained from a study of the form of the zero maximum. If this form were not distorted by "tails" of other interatomic functions, then such knowledge would be absolute. Possibly, in spite of this, the Fourier transform of the zero maximum has a meaning. If we assume that the influence of the other interatomic functions is negligible, then the zero maximum will be represented by the expression

$$\sum \chi_{jj}(\mathbf{R}) = \sum \Phi(f_j^2) = \Phi\left(\sum f_j^2\right);$$

consequently,

$$\sum f_j^2 = \Phi^{-1}\left(\sum \chi_{jj}(\mathbf{R})\right). \tag{7}$$

Selecting a suitable approximation for the distribution of the electron density in the zero maximum, it is possible to obtain the value of $\sum f_j^2$ by computing the Fourier intgral. It should be recalled that experiment gives us P (r) accurately except for an additive constant and without the term $\frac{F^2_{000}}{V}$. However, keeping in mind the good convergence of the series for P (r), a constant magnitude proportional to $\frac{F^2_{000}}{V}$, may be added to the experimental series, so that negative values of P(r) almost disappear.

In any case, the form of the central part of the zero maximum cannot be substantially affected either by the "tails" of other interatomic functions, or by the neglect of the constant term $\frac{F^2_{000}}{V}$. Therefore, the end of the $\overline{f^2}$ curve can be determined with certainty from the Fourier transform of the zero maximum.

3. Picking out the Interatomic Vector System from the Convolution

As we have just seen, the convolution of the electron density represents an aggregate of interatomic functions superimposed one upon another. In contrast with the atomic functions in the series for the electron density, the interatomic functions may approach one another arbitrarily closely; even occasional superpositions are possible. Therefore, one should not expect that all the maxima of a Patterson series will correspond to interatomic functions and thus allow one to determine uniquely the system of interatomic vectors.

The more complex the crystal, the more involved is the picture of the maxima in the series $P(\mathbf{r})$, and the less it corresponds to the system of interatomic vectors of the crystal.

What, then, is the degree of complexity of the crystal when one can no longer speak of the possibility of picking out the interatomic vector system from the series $P(\mathbf{r})$?

The criterion for the possibility of such an isolation is, in principle, the ratio of the number of maxima in the F^2 series to the number of cells in P-function space; if by a cell is meant the volume Δ^3, where Δ is the resolving power of the series and is dependent on the smallest interplanar distance ($\Delta \approx 0.6 d_{min}$) observed. As has been indicated earlier, one rarely can expect $d_{min} < 0.4$ A, and, consequently $\Delta = 0.25$ A at best. As a rule, Δ will be twice as large. If the dimension of the unit cell is a, then the number of cells per lattice constant will be $\frac{a}{\Delta}$, and the number of cells per unit volume of P-function space will be $\left(\frac{a}{\Delta}\right)^3$. If the number of atoms in the cell is denoted by N, then the condition for the possibility of picking out the system of interatomic vectors may be written in the form

$$\left(\frac{a}{\Delta}\right)^3 > N(N-1).$$

The following rough approximation is interesting. A single average atom has the volume $\frac{4}{3}\pi(1.5)^3 A^3 \approx 10-20 A^3$. Instead of a^3 one may substitute 20 N. Then we obtain

$$\frac{20}{\Delta^3} > (N-1).$$

The value Δ in the majority of cases has an order of magnitude of 0.5 A. The condition then acquires the form

$$N < 200.$$

Thus, if there are more than 200 atoms in the unit cell, the possibility of solving the structure problem by current methods becomes small, and, in any case, exceptionally complicated. In these cases, x-ray structure analysis will not serve to find the structure, but to confirm or refute some structural model or other.

This conclusion is quite optimistic, since crystals with $N < 200$ comprise the majority of structures that are of interest for chemistry or mineralogy.

Later, we shall assume that it is possible to pick out the system of interatomic vectors, and set ourselves, the problem of analyzing the method of crystal structure determination when $P(\mathbf{r})$ is known. Let us emphasize that the problem can be solved almost uniquely. The practical complications which we meet are conditioned by only one thing: The overlapping of the interatomic functions often precludes a sufficiently reliable isolation from $P(\mathbf{r})$ of a system of interatomic vectors.

N^2 interatomic vectors can be found in a cell containing N atoms; of these N are trivial ($p = s$), i.e., those connecting an atom "with itself." These N trivial function χ_{ss} constitute the maximum at the point 000. The remaining $N(N-1)$ functions $\chi_{p,s,}$, if their maxima are resolved, give us information on all the interatomic vectors in the structure.

The function $P(\mathbf{r})$ has maxima at those points where the coordinates equal the projections of interatomic vectors. The heights of the maxima are determined by the atomic numbers of the atoms which form the interatomic vector. The number of nontrivial maxima in P-function space is equal to $N(N-1)$.

In a number of cases the symmetry of the crystal may lead to a strict superposition of certain maxima. To have the maxima coincide the crystal must have parallel or antiparallel interatomic vectors.

The only symmetry operation transforming an arbitrary vector into an antiparallel one is the center of inversion. An arbitrary vector is transformed into a parallel one only by translation.

Thus, when a center of inversion is present in the crystal every interatomic vector not passing through the center of symmetry is transformed into an antiparallel one. In the case of a centrosymmetric crystal the interatomic vectors are divided into two groups: N passing through the center of symmetry, and $(N^2/2) - N$ pairs of antiparallel ones. Thus, in P-function space N single and $(N^2/2) - N$ double maxima will be present.

Let us emphasize that this discussion refers to an arbitrary case. It is evident that a vector, either accidentally or due to a parallel symmetry axis or mirror plane, can be transformed into one parallel to itself.

4. The Convolution of the Electron Density and Crystal Symmetry

The function $P(\mathbf{r})$ gives an exhaustive knowledge of the crystal structure (except for the lattice constants and interaxial angles). It defines the symmetry and structure of the crystal.

a) Symmetry of F^2 series and crystal symmetry. The coefficients of the expansion into series of the function $P(\mathbf{r})$ are real and positive magnitudes. In accordance with this, the spatial symmetry of the function $P(\mathbf{r})$ depends on the point group of the crystal. In fact, the equality or inequality of the intensities reflected from the planes hkl comprising one form $\{hkl\}$ characterizes the point group of the crystal, as well as the spatial symmetry of the function $P(\mathbf{r})$. It is not difficult to see that the symmetry elements of the point group of the crystal pass through the point 000 of the space of the function $P(\mathbf{r})$ preserving their arrangement with respect to the origin of coordinates. However, since the intensities of the reflections hkl and $\bar{h}\bar{k}\bar{l}$ are equal, a center of symmetry is added to the aggregate of the symmetry elements of the crystal.

The function $P(\mathbf{r})$ is a periodic function. Consequently, a group of symmetry elements that pass through the point 000 is translated in three dimensions. As to the translation group of the space of $P(\mathbf{r})$, it coincides with the translation group of the crystal.

The summary of the Fedorov space groups of the functions $P(\mathbf{r})$ given in Table 15 originates from what has been said above. The origin of coordinates is always at the center of inversion.

b) Other features of the F^2 series and the symmetry of the crystal. At first glance, it may appear that the F^2 series precludes a judgment of the space group of the crystal, since 230 Fedorov crystal groups are reflected by 24 space groups of the P function.

However, this is not so. The fact is that not only does the symmetry of the P function depend on the symmetry of the crystal, but certain characteristics in the arrangement of the maxima do so also. Each element of symmetry, except the center of inversion, has a corresponding geometric image in the space of the P function.

TABLE 15

Crystal system	Fedorov space group of the crystal	Fedorov space group of P function
Triclinic	Both triclinic groups	P1
Monoclinic	Translation group P	P 2/m
	Translation group C	C2/m
Orthorhombic	Translation group P	P mmm
	Translation group C	C mmm
	Translation group F	F mmm
	Translation group I	I mmm
Tetragonal	Translation group P classes 4, $\bar{4}$ and 4/m	P4/m
	Translation group I classes 4, $\bar{4}$ and 4/m	I4/m
	Translation group P classes $\bar{4}$2 m, 42, 4 mm 4/mmm	P4/mmm
	Translation group I, classes $\bar{4}$2m, 42, 4 mm 4/mmm	I4/mmm
Hexagonal	Translation group C, classes 3 and $\bar{3}$	C3
	Translation group R, classes 3 and $\bar{3}$	R3
	Translation group C, classes 3m, 32, $\bar{3}$m	C3m1 and C31m
	Translation group R, classes 3 m, 32, $\bar{3}$m	R3m
	Translation group C, classes 6, $\bar{6}$, 6/m	C6/m
	Translation group C, classes $\bar{6}$m, 2, 6 mm, 62 and 6/mm	C6/mmm
Cubic	Translation group P, classes 23 and m 3	Pm3
	Translation group F, classes 23 and m 3	Fm3
	Translation group I, classes 23 and m 3	Im3
	Translation group P, classes $\bar{4}$3 m, 43 and m3m	Pm3m
	Translation group,F, classes $\bar{4}$3 m, 43 and m3m	Fm3m
	Translation group I, classes $\bar{4}$3 m, 43 and m3m	Im3m

The presence in the crystal of a rotation axis means the presence of interatomic vectors, the projections of which in the direction of the axis are equal to zero. Thus, rotation axes are reflected in the space of the P function by maxima lying in a plane perpendicular to the axis.

The presence in the crystal of a screw axis means the presence of interatomic vectors the projection of which in the direction of the axis is equal to the screw translation ($^1/_2$ for axis 2_1, $^1/_3$ for axes 3_1 and 3_2, etc.). Thus, screw axes are reflected in the space of the P function by maxima lying in planes perpendicular to the axis and translated by $^1/_2$, $^1/_3$ or $^1/_6$ from the plane containing the origin.

Symmetry planes are reflected in the space of the P function by maxima lying on a line. The presence of a mirror plane of symmetry means the presence of interatomic vectors for which both projections in this plane are equal to zero. Thus, a plane of mirror symmetry is reflected in P-function space by maxima lying along the line through the origin perpendicular to the plane of symmetry. The presence of a glide plane means the presence of interatomic vectors with components $^1/_2$ (or $^1/_4$ in the case of d planes) along the glide axis. Thus, a glide plane is reflected in P-function space by maxima located on a line perpendicular to the glide plane and passing through the middle (or $^1/_4$ for d) of a section of the glide line in the cell. If the case of the plane d is discarded, then lines reflecting glide planes may have the coordinates 0, $^1/_2$; $^1/_2$, 0 or $^1/_2$, $^1/_2$.

From what has been stated, if follows that, in spite the centrosymmetry of the F^2 series, an analysis of it may prove the absence of a center of symmetry in a crystal.

Until recently, it had been assumed that it was impossible to establish the absence or presence of a center of symmetry by an x-ray experiment. As a consequence of this it had been shown that 120 x-ray diffraction groups correspond to the 230 Fedorov groups (it must be remembered that eleven pairs of them are enantiomorphous).

However, with the aid of the F^2series, the number of objectively distinguishable Fedorov groups may be brought to 192. The remaining 38 groups are indistinguishable in pairs. To them belong the eleven enantiomorphous pairs and the eight pairs: I23 and $I2_13$, P3/m and P6/m, R3 and $R\bar{3}$, P3 and $P\bar{3}$, I4 and $I\bar{4}$, P4 and $P\bar{4}$, I222 and $I2_12_12_1$, and P1 and $P\bar{1}$.

If, for example, one were to compare the groups Pa and P2/a (C_s^4 and (C_{2h}^4), which differ from one another only by a center of inversion (and, consequently, are indistinguishable by interference extinctions), then the P-function

TABLE 1. The Characteristics of P-Function Space for the 230 Fedorov Space Groups

Crystal system	Fedorov space group of the crystal	Fedorov space group of P-function space	Images of symmetry planes in P-function space			Images of symmetry axes in P-function space			
Triclinic	$P1$	$P1$	} —	—	—	—	—	—	—
	$P1$								
Monoclinic	$P2$	$P2/m$	—	—	—	—	—	$x0z$	—
	$P2_1$		—	—	—	—	—	$x{}^1/_2z$	—
	Pm		—	—	$0y0$	—	—	—	—
	Pc		—	—	$0y{}^1/_2$	—	—	—	—
	—		—	—	$0y0$	—	—	$x0z$	—
	$P2/c$		—	—	$0y{}^1/_2$	—	—	$x0z$	—
			—	—	$0y0$	—	—	$x{}^1/_2z$	—
	$P2/c$		—	—	$0y{}^1/_2$	—	—	$x{}^1/_2z$	—
	$C2$	$C2/m$	—	—	—	—	—	$x0z$	—
	Cm		—	—	$0y0$	—	—	—	—
	Cc		—	—	$0y{}^1/_2$	—	—	—	—
	$C2/m$		—	—	$0y0$	—	—	$x0z$	—
	$C2/c$		—	—	${}^1/_2$	—	—	$x0z$	—

TABLE 16 (continued)

Crystal system	Fedorov space group of the crystal	Fedorov space group of P-function space	Images of symmetry planes in P-function space			Images of symmetry axes in P-function space			
Orthorhombic	$P222$	$Pmmm$	—	—	—	$0yz$	$x0z$	$xy0$	—
	$P222_1$	—	—	—	—	$0yz$	$x0z$	$xy\frac{1}{2}$	—
	$P2_12_12_1$	—	—	—	—	$\frac{1}{2}yz$	$x\frac{2}{2}z$	$xy0$	—
	$P2_12_12_1$	—	—	—	—	$\frac{1}{2}yz$	$x\frac{1}{2}z$	$xy\frac{1}{2}$	—
	$Pmm2$	—	$x00_1$	$0y0_2$	—	—	—	$xy0_{1,2}$	—
	$Pmc2_1$	—	$x00$	$0y\frac{1}{2}_1$	—	—	—	$xy\frac{1}{2}_1$	—
	$Pcc2$	—	$x0\frac{1}{2}$	$0y\frac{1}{2}$	—	—	—	$xy0$	—
	$Pma2$	—	$x00_1$	$\frac{1}{2}y0_2$	—	—	—	$xy0_{1,2}$	—
	$Pca2_1$	—	$x0\frac{1}{2}_1$	$\frac{1}{2}y0$	—	—	—	$xy\frac{1}{2}_1$	—
	$Pnc2$	—	$x\frac{1}{2}\frac{1}{2}$	$0y\frac{1}{2}$	—	—	—	$xy0$	—
	$Pmn2_1$	—	$x00$	$\frac{1}{2}y\frac{1}{2}_1$	—	—	—	$xy\frac{1}{2}_1$	—
	$Pba2$	—	$x\frac{1}{2}0_1$	$\frac{1}{2}y0_2$	—	—	—	$xy0_{1,2}$	—
	$Pna2_1$	—	$x\frac{1}{2}\frac{1}{2}_1$	$\frac{1}{2}y0$	—	—	—	$xy\frac{1}{2}_1$	—
	$Pnn2$	—	$x\frac{1}{2}\frac{1}{2}$	$\frac{1}{2}y\frac{1}{2}$	—	—	—	$xy0$	—
	$Pmmm$		$x00_{1,2}$	$0y0_{3,4}$	$00z_{5,6}$	$0yz_{3,5}$	$x0z_{1,6}$	$xy0_{2,4}$	–
	$Pnnn$	—	$x\frac{1}{2}\frac{1}{2}$	$\frac{1}{2}y\frac{1}{2}$	$\frac{1}{2}\frac{1}{2}z$	$0yz$	$x0z$	$xy0$	–

TABLE 16 (continued)

Crystal system	Fedorov space group of the crystal	Fedorov space group of P-function space	Images of symmetry planes in P-function space			Images of symmetry axes in P-function space			
Orthorhombic	*Pccm*	—	$x0\frac{1}{2}_1$	$0y\frac{1}{2}_2$	$00z_{3,4}$	$0yz_{2,3}$	$x0z_{1,4}$	$xy0$	—
	Pban	—	$x\frac{1}{2}0_1$	$\frac{1}{2}y0_2$	$\frac{1}{2}\frac{1}{2}z$	$0yz$	$x0z$	$xy0_{1,2}$	—
	Pmma	—	$x00_{1,2}$	$0y0_3$	$\frac{1}{2}0z_{4,5}$	$\frac{1}{2}yz_4$	$x0z_{1,5}$	$xy0_{2,3}$	—
	Pnna	—	$x\frac{1}{2}\frac{1}{2}_1$	$\frac{1}{2}y\frac{1}{2}$	$\frac{1}{2}0z$	$0yz$	$x\frac{1}{2}z_1$	$xy0$	—
	Pmna	—	$x00_1$	$\frac{1}{2}y\frac{1}{2}_2$	$\frac{1}{2}0z_3$	$0yz$	$x0z_{1,3}$	$xy\frac{1}{2}_2$	—
	Pcca	—	$x0\frac{1}{2}0_1$	$0y\frac{1}{2}$	$\frac{1}{2}0z_{2,3}$	$\frac{1}{2}yz_2$	$x0z_{1,3}$	$xy0$	—
	Pbam	—	$x\frac{1}{2}0_1$	$\frac{1}{2}y0_{2,3}$	$00z_4$	$\frac{1}{2}yz_2$	$x0z_4$	$xy0_{1,3}$	—
	Pccn	—	$x0\frac{1}{2}$	$0y\frac{1}{2}$	$\frac{1}{2}\frac{1}{2}z_{1,2}$	$\frac{1}{2}yz_1$	$x\frac{1}{2}z_2$	$xy0$	—
	Pbcm	—	$x\frac{1}{2}0_1$	$0y\frac{1}{2}_{2,3}$	$00z_4$	$0yz_{2,4}$	$x\frac{1}{2}z_1$	$xy\frac{1}{2}_3$	—
	Pnnm	—	$x\frac{1}{2}\frac{1}{2}_1$	$\frac{1}{2}y\frac{1}{2}_2$	$00z$	$\frac{1}{2}yz_2$	$x\frac{1}{2}z_1$	$xy0$	—
	Pmmn	—	$x00_1$	$0y0_2$	$\frac{1}{2}\frac{1}{2}z_{3,4}$	$\frac{1}{2}yz_2$	$x\frac{1}{2}z_4$	$xy0_{1,2}$	—
	Pbcn	—	$x\frac{1}{2}0$	$0y\frac{1}{2}_1$	$\frac{1}{2}\frac{1}{2}z_2$	$\frac{1}{2}yz_2$	$x0z$	$xy\frac{1}{2}_1$	—
	Pbca	—	$x\frac{1}{2}0_1$	$0y\frac{1}{2}_3$	$\frac{1}{2}0z_3$	$\frac{1}{2}yz_3$	$x\frac{1}{2}z_1$	$x0\frac{1}{2}_2$	—
	Pnma	—	$x\frac{1}{2}\frac{1}{2}_{1,2}$	$0y0$	$\frac{1}{2}0z_3$	$\frac{1}{2}yz_3$	$x\frac{1}{2}z_1$	$x0\frac{1}{2}_2$	—
	$C222_1$	C	—	—	—	$0yz$	$x0z$	$xy\frac{1}{2}$	—
	$C222$	—	—	—	—	$0yz$	$x0z$	$xy0$	—

TABLE 16 (continued)

Crystal system	Fedorov space group of the crystal	Fedorov space group of P-function space	Images of symmetry planes in P-function space			Images of symmetry axes in P-function space			
Orthorhombic	$Cmm2$	—	$x00_1$	$0y0_2$	—	—	—	$xy0_{1,2}$	—
	$Cmc2_1$	—	$x00$	$0y^1/_{2\,1}$		—	—	$xy^1/_{2\,1}$	—
	$Ccc2$	—	$x0^1/_2$	$0y^1/_2$	—	—	—	$xy0$	—
		—	—	$0y0_1$	$00z_2$	$0yz_{1,2}$	—	—	—
		—	—	$0y0_1$	$0^1/_2z_2$	$0yz_{1,2}$	—	—	—
		—	—	$0y^1/_{2\,1}$	$00z_2$	$0yz_{1,2}$	—	—	—
		—	—	$0y^1/_{2\,1}$	$0^1/_2z_2$	$0yz_{1,2}$	—	—	—
	$Cmcm$	—	$x00_1$	$0y^1/_{2\,2,3}$	$00z_{4,5}$	$0yz_{2,4}$	$x0z_{1,5}$	$xy^1/_{2\,3}$	—
	$Cmca$	—	$x00_1$	$0y^1/_{2\,2,3}$	$^1/_20z_{4,5}$	$0yz_{2,4}$	$x0z_{1,5}$	$xy^1/_{2\,3}$	—
	$Cmmm$	—	$x00_{1,2}$	$0y0_{3,4}$	$00z_{5,6}$	$0yz_{3,5}$	$x0z_{1,6}$	$xy0_{2,4}$	—
	$Cccm$	-	$x0^1/_{2\,1}$	$0y^1/_{2\,2}$	$00z_{3,4}$	$0yz_{2,3}$	$x0z_{1,4}$	$xy0$	—
	$Cmma$	—	$x00_{1,2}$	$0y0_{3,4}$	$^1/_20z_5$	$0yz_{3,5}$	$x0z_1$	$xy0_{2,4}$	—
	$Ccca$	—	$x0^1/_{2\,1}$	$0y^1/_{2\,2}$	$^1/_20z_{3,5}$	$0yz_{2,5}$	$x0z_{1,3}$	$xy0$	—
	$I222$ $I2_12_12_1$	$Immm$	} —	—	—	$0yz$	$x0z$	$xy0$	—

TABLE 16 (continued)

Crystal system	Fedorov space group of the crystal	Fedorov space group of P-function space	Images of symmetry planes in P-function space			Images of symmetry axes in P-function space			
Orthorhombic	*Imm*2	—	$x00_1$	$0y0_2$	—	—	—	$xy0_{1,2}$	—
	*Iba*2	—	$x\frac{1}{2}0_1$	$\frac{1}{2}y0_2$	—	—	—	$xy0_{1,2}$	—
	*Ima*2	—	$x00_1$	$\frac{1}{2}y0_2$	—	—	—	$xy0_{1,2}$	—
	Immm	—	$x00_{1,2}$	$0y0_{3,4}$	$00z_{5,6}$	$0yz_{3,5}$	$x0z_{1,6}$	$xy0_{2,4}$	—
	Ibam	—	$x\frac{1}{2}0_{1,2}$	$\frac{1}{2}y0_{3,4}$	$00z_{5,6}$	$0yz_{3,5}$	$x0z_{1,6}$	$xy0_{2,4}$	—
	Ibca	—	$x\frac{1}{2}0_{1,2}$	$0y\frac{1}{2}_{3,4}$	$\frac{1}{2}0z_{5,5}$	$0yz_{3,5}$	$x0z_{1,6}$	$xy0_{2,4}$	—
	Imma	—	$x00_{1,2}$	$0y0_{3,4}$	$\frac{1}{2}0z_{5,6}$	$0yz_{3,5}$	$x0z_{1,6}$	$xy0_{2,4}$	—
	*F*222	*Fmmm*	—	—	—	$0yz$	$x0z$	$xy0$	—
	*Fmm*2	—	$x00_1$	$0y0_2$	—	—	—	$xy0_{1,2}$	—
	*Fdd*2	—	$x\frac{1}{4}\frac{1}{4}$	$\frac{1}{4}y\frac{1}{4}$	—	—	—	$xy0$	—
	Fmmm	—	$x00_{1,2}$	$0y0_{3,4}$	$00z_{5,6}$	$0yz_{3,5}$	$x0z_{1,6}$	$xy0_{2,4}$	—
	Fddd	—	$x\frac{1}{4}\frac{1}{4}$	$\frac{1}{4}y\frac{1}{4}$	$\frac{1}{4}\frac{1}{4}z$	$0yz$	$x0z$	$xy0$	—

TABLE 16 (continued)

Crystal system	Fedorov space group of the crystal	Fedorov space group of P-function space	Images of symmetry planes in P-function space			Images of symmetry axes in P-function space			
Tetragonal	$P4$	$P4/m$	—	—	—	—	—	$xy0$	—
	$P4$	—	—	—	—	—	—	—	—
	$P4_1$	—	—	—	—	—	—	—	—
	$P4_3$	—	—	—	—	—	—	$xy^1/_4$	$xy^1/_2$
	$P4_2$	—	—	—	—	—	—	$xy0$	$xy^1/_2$
	$P4/m$	—	—	—	$00z$	—	—	$xy0$	—
	$P4_2/m$	—	—	—	$00z$	—	—	$xy0$	$xy^1/_2$
	$P4/n$	—	—	—	$^1/_2\,^1/_2\,z$	—	—	$xy0$	—
	$P4_2/n$	—	—	—	$^1/_2\,^1/_2\,z$	—	—	$xy0$	$xy^1/_2$
	$I\bar{4}$ }	$I4/m$	—	—	—	—	—	$xy0$	—
	$I4$ }	—							
	$I4_1$	—	—	—	—	—	—	$xy0$	$xy^1/_4$
	$I4/m$	—	—	—	$00z$	—	—	$xy0$	—
	$I4_1/a$	—	—	—	$^1/_2 0 z$	—	—	$xy0$	$xy^1/_4$

TABLE 16 (continued)

Crystal system	Fedorov space group of the crystal	Fedorov space group of P-function space	Images of symmetry planes in P-function space			Images of symmetry axes in P-function space			
Tetragonal	$P422$	$P4/mmm$	—	—	—	$x0z$	xxz	$xy0$	—
	$P42_12$	—	—	—	—	$x^1/_2z$	xxz	$xy0$	—
	$P4_122$	—	—	—	—	$x0z$	xxz	$xy^1/_4$	$xy^1/_2$
	$P4_322$	—							
	$P4_12_12$	—	—	—	—	$x^1/_2z$	xxz	$xy^1/_4$	$xy^1/_2$
	$P4_32_12$	—							
	$P4_222$	—	—	—	—	$x0z$	xxz	$xy0$	$xy^1/_2$
	$P4_22_12$	—	—	—	—	$x^2/_2z$	xxz	$xy0$	$xy^1/_2$
	$P4mm$	—	$x00_1$	$xx0_2$	—	—	—	$xy0_{1,2}$	—
	$P4bn$	—	$x^1/_20_1$	$xx0_2$	—	—	—	$xy0_{1,2}$	—
	$P4_2cm$	—	$x0^1/_{2_1}$	$xx0_2$	—	—	—	$xy0_2$	$xy^1/_{2_1}$
	$P4_2nm$	—	$x^1/_2{}^1/_{2_1}$	$xx0$	—	—	—	$xy0_2$	$xy^1/_{2_1}$
	$P4cc$	—	$x0^1/_2$	$xx^1/_2$	—	—	—	$xy0$	—
	$P4nc$	—	$x^1/_2{}^1/_2$	$xx^1/_2$	—	—	—	$xy0$	—
	$P4_2mc$	—	$x00_1$	$xx^1/_{2_2}$	—	—	—	$xy0_1$	$xy^1/_{2_2}$
	$P4_2bc$	—	$x^1/_20_1$	$xx^1/_{2_2}$	—	—	—	$xy0_1$	$xy^1/_{2_2}$

TABLE 16 (continued)

Crystal system	Fedorov space group of the crystal	Fedorov space group of P-function space	Images of symmetry planes in P-function space			Images of symmetry axes in P-function space			
Tetragonal	$P\bar{4}2m$	—	—	$xx0_1$	—	$x0z$	—	$xy0_1$	—
	$P\bar{4}2c$	—	—	$xx\frac{1}{2}$	—	$x0z$	—	$xy0$	—
	$P\bar{4}2_1m$	—	—	$xx0_1$	—	$x\frac{1}{2}z$	—	$xy0_1$	—
	$P\bar{4}2_1c$	—	—	$xx\frac{1}{2}$	—	$x\frac{1}{2}z$	—	$xy0$	—
	$P\bar{4}m2$	—	$x00_1$	—	—	—	xxz	$xy0_1$	—
	$P\bar{4}c2$	—	$x0\frac{1}{2}$	—	—	—	xxz	$xy0$	—
	$P\bar{4}b2$	—	$x\frac{1}{2}0_1$	—	—	—	xxz	$xy0_1$	—
	$P\bar{4}n2$	—	$x\frac{1}{2}\frac{1}{2}$	—	—	—	xxz	$xy0$	—
	$P4/mmm$	—	$x00_{1,2}$	$xx0_{3,4}$	$00z_{5,6}$	$x0z_{1,5}$	$xxz_{3,6}$	$xy0_{2,4}$	—
	$P4/mcc$	—	$x0\frac{1}{2}_1$	$xx\frac{1}{2}_2$	$00z_{3,4}$	$x0z_{1,3}$	$xxz_{2,4}$	$xy0$	—
	$P4/nbm$	—	$x\frac{1}{2}0_1$	$xx0_{2,3}$	$\frac{1}{2}\frac{1}{2}z_4$	$x0z$	$xxz_{2,4}$	$xy0_{1,3}$	—
	$P4/nnc$	—	$x\frac{1}{2}\frac{1}{2}$	$xx\frac{1}{2}_1$	$\frac{1}{2}\frac{1}{2}z_2$	$x0z$	$xxz_{1,2}$	$xy0$	—
	$P4/mbm$	—	$x\frac{1}{2}0_{1,2}$	$xx0_{3,4}$	$00z_5$	$x\frac{1}{2}z_1$	$xxz_{3,5}$	$xy0_{2,4}$	—
	$P4/mnc$	—	$x\frac{1}{2}\frac{1}{2}_1$	$xx\frac{1}{2}_2$	$00z_3$	$x\frac{1}{2}z_1$	$xxz_{2,3}$	$xy0$	—
	$P4/nmm$	—	$x00_1$	$xx0_{2,3}$	$\frac{1}{2}\frac{1}{2}_{4,5}$	$x\frac{1}{2}z_4$	$xxz_{2,5}$	$xy0_{1,3}$	—
	$P4/ncc$	—	$x0\frac{1}{2}$	$xx\frac{1}{2}_1$	$\frac{1}{2}\frac{1}{2}_{2,3}$	$x\frac{1}{2}z_2$	$xxz_{1,3}$	$xy0$	—
	$P4_2/mmc$	—	$x00_{1,2}$	$xx0_{3,4}$	$00z_{5,6}$	$x0z_{1,5}$	$xxz_{3,6}$	$xy0_{2,4}$	$xy\frac{1}{2}$

TABLE 16 (continued)

Crystal system	Fedorov space group of the crystal	Fedorov space group of P-function space	Images of symmetry planes in P-function space			Images of symmetry axes in P-function space			
Tetragonal	$P4_2/mcm$	—	$x0\tfrac12_1$	$xx0_{2,3}$	$00z_{4,5}$	$x0z_4$	$xxz_{2,5}$	$xy0_3$	$xy\tfrac12_1$
	$P4_2/nbc$	—	$x\tfrac12 0_1$	$xx\tfrac12_{2,3}$	$\tfrac12\tfrac12 z_4$	$x0z$	$xxz_{2,4}$	$xy0_1$	$xy\tfrac12_3$
	$P4_2/nnm$	—	$x\tfrac12\tfrac12_1$	$xx0_{2,3}$	$\tfrac12\tfrac12 z_4$	$x0z$	$xxz_{2,4}$	$xy0_3$	$xy\tfrac12_1$
	$P4_2/mbc$	—	$x\tfrac12 0_{1,2}$	$xx\tfrac12_{3,4}$	$00z_5$	$x\tfrac12 z_1$	$xxz_{3,5}$	$xy0_2$	$xy\tfrac12_4$
	$P4_2/mnm$	—	$x\tfrac12\tfrac12_{1,2}$	$xx0_{3,4}$	$00z_5$	$x\tfrac12 z_1$	$xxz_{3,5}$	$xy0_4$	$xy\tfrac12_2$
	$P4_2/nmc$	—	$x00_1$	$x\tfrac12_{2,3}$	$\tfrac12\tfrac12 z_{5,6}$	$x\tfrac12 z_5$	$xxz_{2,6}$	$xy0_1$	$xy\tfrac12_3$
	$P4_2/ncm$	—	$x0\tfrac12_1$	$x0_{2,3}$	$\tfrac12\tfrac12 z_{5,6}$	$x\tfrac12 z_5$	$xxz_{2,6}$	$xy0_3$	$xy\tfrac12_1$
	$I422$	$I4/mmm$	—	—	—	$x0z$	xxz	$xy0$	—
	$I4_122$	—	—	—	—	$x0z$	xxz	$xy0$	$xy\tfrac14$
	$I4mm$	—	$x00_1$	$xx0_2$	—	—	—	$xy0_{1,2}$	—
	$I4cm$	—	$x0\tfrac12_1$	$xx0_2$	—	—	—	$xy0_{1,2}$	—
	$I4_1md$	—	$x00_1$	$x/x+\tfrac12/\tfrac14{}_2$	—	—	—	$xy0_1$	$xy\tfrac14{}_2$
	$I4_1cd$	—	$x0\tfrac12_1$	$x/x+\tfrac12/\tfrac14{}_2$	—	—	—	$xy0_1$	$xy\tfrac14{}_2$
	$I\bar{4}m2$	—	$x00_1$	—	—	—	xxz	$xy0_1$	—
	$I\bar{4}c2$	—	$x0\tfrac12_1$	—	—	—	xxz	$xy0_1$	

TABLE 16 (continued)

Crystal system	Fedorov space group of the crystal	Fedorov space group of P-function space	Images of symmetry planes in P-function space			Images of symmetry axes in P-function space			
Tetragonal	$I\bar{4}2m$	—	—	$xx0_1$	—	$x0^z$	—	$xy0_1$	—
	$I\bar{4}2d$	—	—	$x/x+{}^1/_2/{}^1/_4$	—	$x0z$	—	$xy0$	—
	$I4/mmm$	—	$x00_{1,2}$	$xx0_{3,4}$	$00z_{5,6}$	$x0z_{1,5}$	$xxz_{3,6}$	$xy0_{2,4}$	—
	$I4/mcm$	—	$x0{}^1/_2{}_{1,2}$	$xx0_{3,4}$	$00z_{5,6}$	$x0z_{1,5}$	$xxz_{3,6}$	$xy0_{2,4}$	—
	$I4/amd$	—	$x00_{1,2}$	$x/x+{}^1/_2/{}^1/_4{}_3$	${}^1/_2 0z_4$	$x0z_{1,4}$	xxz	$xy0_2$	$xy{}^1/_4{}_2$
	$I4/acd$	—	$x0{}^1/_2{}_{1,2}$	$x/x+{}^1/_2/{}^1/_4{}_3$	${}^1/_2 0z_4$	$x0z_{1,4}$	xxz	$xy0_2$	$xy{}^1/_4{}_3$
Hexagonal	$P\bar{3}$	$P\bar{3}$	} —	—	—	—	—	$xy0$	—
	$P3$	—							
	$P3_1$	—	} —	—	—	—	—	$xy^1/_3$	—
	$P3_1$	—							
	$R3$	$R\bar{3}$	} —	—	—	—	—	$xy0$	—
	$R\bar{3}$								
	$P321$	$P\bar{3}m1$	—	—	—	$x2xz$	—	$xy0$	—
	$P3_121$		} —	—	—	$x2xz$	—	$xy^1/_3$	—
	$P3_221$								

TABLE 16 (continued)

Crystal system	Fedorov space group of the crystal	Fedorov space group of P-function space	Images of symmetry planes in P-function space			Images of symmetry axes in P-function space			
Hexagonal	$P3m1$	—	—	$xx0_1$	—	—	—	$xy0_1$	—
	$P3c1$	—	—	$xx^1/_2$	—	—	—	$xy0$	—
	$P\bar{3}m1$	—	—	$xx0_1$	—	$x2xz$	—	$xy0_1$	—
	$P\bar{3}c1$	—	—	$xx^1/_2$	—	$x2xz$	—	$xy0$	—
	$P312$	$P\bar{3}1m$	—	—	—	—	xxz	$xy0$	—
	$P3_412$	—	—	—	—	—	xxz	$xy^1/_3$	—
	$P3_212$	—							
	$P31m$	—	$x2x0_1$	—	—	—	—	$xy0_1$	—
	$P31c$	—	$x2x^1/_2$	—	—	—	—	$xy0$	—
	$P\bar{3}1m$	—	$x2x0_1$	—	—	—	xxz	$xy0_1$	—
	$P31c$	—	$x2x^1/_2$	—	—	—	xxz	$xy0$	—
	$R32$	$R3m$	—	—	—	$x2xz$	—	$xy0$	—
	$R3m$	—	—	$xx0_1$	—	—	—	$xy0_1$	—
	$R3c$	—	—	$xx^1/_2$	—	—	—	$xy0$	—

TABLE 16 (continued)

Crystal system	Fedorov space group of the crystal	Fedorov space group of P-function space	Images of symmetry planes in P-function space			Images of symmetry axes in P-function space				
Hexagonal	$R\bar{3}m$	—	—	$xx0_1$	—	$x2xz$	—	$xy0_1$	—	—
	$R\bar{3}c$	—	—	$xx^{1}/_{2}$	—	$x2xz$	—	$xy0$	—	—
	$P6$	$P6/m$	—	—	—	—	—	$xy0$	—	—
	$P6_1$	—	—	—	—	—	—	$xy^{1}/_{6}$	$xy^{1}/_{3}$	$xy^{1}/_{2}$
	$P6_5$	—								
	$P6_2$	—	—	—	—	—	—	$xy0$	$xy^{1}/_{3}$	—
	$P6_4$	—								
	$P6_3$	—	—	—	—	—	—	$xy0$	—	$xy^{1}/_{2}$
	$P\bar{6} = P3/m$	—	—	—	$00z$	—	—	$xy0$	—	—
	$P6/m$	—								
	$P6_3/m$	—	—	—	$00z$	—	—	$xy0$	—	$xy^{1}/_{2}$
	$P622$	$P6/mmm$	—	—	—	$x2xz$	xxz	$xy0$	—	—
	$P6_122$		—	—	—	$x2xz$	xxz	$xy^{1}/_{6}$	$xy^{1}/_{3}$	$xy^{1}/_{2}$
	$P6_522$									

TABLE 16 (continued)

Crystal system	Fedorov space group of the crystal	Fedorov space group of P-function space	Images of symmetry planes in P-function space			Images of symmetry axes in P-function space				
Hexagonal	$P6_222$	—								
	$P6_422$	—	—	—	—	$x2xz$	xxz	$xy0$	$xy{}^1/_3$	—
	$P6_322$	—	—	—	—	$x2xz$	xxz	$xy0$	—	$xy{}^1/_2$
	$P6mm$	—	$x2x0_1$	$xx0_2$	—	—	—	$xy0_{1,2}$	—	—
	$P6cc$	—	$x2x{}^1/_2$	$xx{}^1/_2$	—	—	—	$xy0$	—	—
	$P6_3cm$	—	$x2x0_1$	$xx{}^1/_{2_2}$	—	—	—	$xy0_1$	—	$xy{}^1/_{2_2}$
	$P6_3mc$	—	$x2x{}^1/_{2_1}$	$xx0_2$	—	—	—	$xy0_2$	—	$xy{}^1/_{2_1}$
	$P6m2$	—	—	$xx0_{1,2}$	—	—	xxz_1	$xy0_3$	—	—
	$P\bar{6}c2$	—	—	$xx{}^1/_{2_1}$	—	—	xxz_1	$xy0$	—	—
	$P\bar{6}2m$	—	$x2x0_{1,2}$	—	—	$x2xz_1$	—	$xy0_2$	—	—
	$P\bar{6}2c$	—	$x2x{}^1/_{2_1}$	—	—	$x2xz_1$	—	$xy0$	—	—
	$P6/mmm$	—	$x2x0_{1,2}$	$xx0_{3,4}$	$00z_{5,6}$	$x2xz_{1,5}$	$xxz_{3,6}$	$xy0_{2,4}$	—	—
	$P6/mcc$	—	$x2x{}^1/_{2_1}$	$xx{}^1/_{2_2}$	$00z_{3,4}$	$x2xz_{1,3}$	$xxz_{2,4}$	$xy0$	—	—
	$P6_3/mcm$	—	$x2x0_{1,2}$	$xx{}^1/_{2_{3,4}}$	$00z_{5,6}$	$x2xz_{1,5}$	$xxz_{3,6}$	$xy0_2$	—	$xy{}^1/_{2_4}$
	$P6_3/mmc$	—	$x2x{}^1/_{2_{1,2}}$	$xx0_{3,4}$	$00z_{5,6}$	$x2xz_{1,5}$	$xxz_{3,6}$	$xy0_4$	—	$xy{}^1/_{2_2}$

TABLE 16 (continued)

Crystal system	Fedorov space group of the crystal	Fedorov space group of P-function space	Images of symmetry planes in P-function space			Images of symmetry axes in P-function space				
Cubic	$P23$	$Pm3$	—	—	—	—	$xy0$	—	$xy(x+y)$	—
	$P2_13$	—	—	—	—	—	$xy\frac{1}{2}$	—	$xy(x+y)$	—
	$Pm3$	—	$x00_1$	—	—	—	$xy0_1$	—	$xy(x+y)$	—
	$Pn3$	—	$x\frac{1}{2}\frac{1}{2}$	—	—	—	$xy0$	—	$xy(x+y)$	—
	$Pa3$	—	$x\frac{1}{2}0_1$	—	—	—	$xy\frac{1}{2}{}_1$	—	$xy(x+y)$	—
	$I23$	$Im3$	} —	—	—	—	$xy0$	—	$xy(x+y)$	—
	$I2_13$	—								
	$Im3$	—	$x00_1$	—	—	—	$xy0_1$	—	$xy(x+y)$	—
	$Ia3$	—	$x\frac{1}{2}0_1$	—	—	—	$xy0_1$	—	$xy(x+y)$	—
	$F23$	$Fm3$	—	—	—	—	$xy0$	—	$xy(x+y)$	—
	$Fm3$	—	$x00_1$	—	—	—	$xy0_1$	—	$xy(x+y)$	—
	$Fd3$	—	$x\frac{1}{4}\frac{1}{4}$	—	—	—	$xy0$	—	$xy(x+y)$	—
	$P432$	$Pm3m$	—	—	—	xxz	$xy0$	—	$xy(x+y)$	—
	$P4_232$	—	—	—	—	xxz	$xy0$	$xy\frac{1}{2}$	$xy(x+y)$	—

TABLE 16 (continued)

Crystal system	Fedorov group of the crystal	Fedorov space group of P-function space	Images of symmetry planes in P-function space			Images of symmetry axes in P-function space				
Cubic	$P4_332$	—	} —	—	—	xxz	$xy^{1/4}$	$xy^{1/2}$	$xy(x+y)$	—
	$P4_132$	—								
	$P\bar{4}3m$	—	—	$xx0_1$	—	—	$xy0_1$	—	$xy(x+y)$	—
	$P\bar{4}3n$	—	—	$xx^{1/2}$	—	—	$xy0$	—	$xy(x+y)$	—
	$Pm3m$	—	$x00_{1,2}$	$xx0_{3,4}$	—	$xxz_{1,3}$	$xy0_{2,4}$	—	$xy(x+y)$	—
	$Pn3n$	—	$x^{1/2}\,{}^{1/2}{}_1$	$xx^{1/2}{}_2$	—	$xxz_{1,2}$	$xy0$	—	$xy(x+y)$	—
	$Pn3n$	—	$x00_{1,2}$	$xx^{1/2}{}_{3,4}$	—	$xxz_{1,3}$	$xy0_2$	$xy^{1/2}{}_4$	$xy(x+y)$	—
	$Pn3m$	—	$x^{1/2}\,{}^{1/2}{}_{1,2}$	$xx0_{3,4}$	—	$xxz_{1,3}$	$xy0_4$	$xy^{1/2}{}_2$	$xy(x+y)$	—
	$I432$	$Im3m$	—	—	—	xxz	$xy0$	—	$xy(x+y)$	—
	$I4_132$	—	—	—	—	xxz	$xy0$	$xy^{1/4}$	$xy(x+y)$	—
	$I\bar{4}3m$	—	—	$xx0_1$	—	—	$xy0_1$	—	$xy(x+y)$	—
	$I\bar{4}3d$	—	—	$x(x+{}^1/_2)^{1/4}$	—	—	$xy0$	—	$xy(x+y)$	—
	$Im3m$	—	$x00_{1,2}$	$xx0_{3,4}$	—	$xxz_{1,3}$	$xy0_{2,4}$	—	$xy(x+y)$	—
	$Ia3d$	—	$x^{1/2}0_1$	$x(x+{}^1/_2)^{1/4}{}_2$	—	xxz	$xy0_1$	$xy^{1/4}{}_2$	$xy(x+y)$	—

TABLE 16 (continued)

Crystal system	Fedorov space group of the crystal	Fedorov space group of P-function space	Images of symmetry planes in P-function space			Images of symmetry axes in P-function space				
Cubic	$F\bar{4}32$	$Fm3m$	—	—	—	xxz	$xy0$	—	$xy(x+y)$	—
	$F\bar{4}_132$	—	—	—	—	xxz	$xy0$	$xy^1/_4$	$xy(x+y)$	—
	$F\bar{4}3m$	—	—	$xx0_{1,2}$	—	xxz_1	$xy0_2$	—	$xy(x+y)$	—
	$F\bar{4}3c$	—	—	$xx^1/_{2\,1,2}$	—	xxz_1	$xy0_2$	—	$xy(x+y)$	—
	$Fm3m$	—	$x00_{1,2}$	$xx0_{3,4}$	—	$xxz_{1,3}$	$xy0_{2,4}$	—	$xy(x+y)$	—
	$Fm3c$	—	$x00_{1,2}$	$xx^1/_{2\,3,4}$	—	$xxz_{1,3}$	$xy0_{2,4}$	—	$xy(x+y)$	—
	$Fd3m$	—	$x^1/_4{}^1/_{4\,1,2}$	$xx0_{3,4}$	—	$xxz_{1,3}$	$xy0_4$	$xy^1/_{4\,2}$	$xy(x+y)$	—
	$Fd3c$	—	$x^1/_4{}^1/_{4\,1,2}$	$xx^1/_{2\,3,4}$	—	$xxz_{1,3}$	$xy0_4$	$xy^1/_{4\,2}$	$xy(x+y)$	—

Note: This table is constructed on the basis of principles stated in the text. The name of the crystal system is given in the first column. In the second column, the symbol and the orientation of the Fedorov space group are given. In the third column, the Fedorov space group of the P function is given.

In the next column, the maxima of the F^2 series lying on lines parallel to the axes of coordinates are given. These maxima are images of symmetry planes perpendicular to the axes x, y, and z in the accepted orientation of the Fedorov space group. Only the independent maxima are given; the remaining ones may be obtained from these by the application of the symmetry elements of the Fedorov space group of the P function.

The last column gives the images of symmetry axes. Here also, only the independent maxima of the F^2 series are given. Space groups indistinguishable by the distribution of the maxima in the F^2 series are singled out by parentheses. This table was constructed by M. J. Buerger.

Subscripts are appended to the coordinate triples to show which two dimensional sections in the last column contain one dimensional sections in the penultimate column. The aggregate of the data given in the two last columns is nothing but a summary of the symmetrically independent maxima in P-function space created by a single atom in a general position. For example, if in the Fedorov space group Pbam there is one atom in a general position (i.e., since the multiplicity of the group is eight, there is a total of eight atoms in the cell), then the F^2 series will give maxima on the lines $x\frac{1}{2}0$, $\frac{1}{2}y0$ and $00z$, on the planes, $\frac{1}{2}yz$, $x0z$ and $xy0$, and also, of course, a maximum in a general position.

It is not difficult to calculate the relative heights of the maxima. There are $8 \times 7 = 56$ interatomic vectors in all, and thus, there are seven maxima in the asymmetric unit of P-function space. Of them one is in a general position corresponding to the vector connecting the centrosymmetrically rotated atoms. Let us consider the height of this maximum to be one unit. Atoms connected by symmetry planes given an interatomic vector parallel to the two other symmetry planes. There are four such parallel vectors. Consequently, the maxima on the lines will have a height of four lines; they are like a quadruplet of merged maxima. There is $\frac{1}{4}$ of each such merged maximum in the asymmetric unit of P-function space.

Maxima on the planes have a height of two units. Each maximum is formed by two interatomic vectors.

The description of the F^2 series given in the table is not complete. For a special position, a method of describing the P-function space should be used that corresponds to the Fedorov space group of the crystal: The Fedorov group is characterized by several distributions of points (several special and one general) a, b, c, d,..., each position, if it is occupied by atoms, gives in P-function space a typical combination of maxima aa if only one such position is occupied and, in addition, ab, ac, ad,..., if other positions are also occupied by atoms.

Consequently, the P-function space corresponding to the Fedorov group with, for example, four possible positions a, b, c, d, should be characterized by giving ten types of distributions by maxima:

aa	*ab*	*ac*	*ad*
	bb	*bc*	*bd*
		cc	*cd*
			dd

Relative heights of maxima are also missing from the table above.

spaces should appear different. The first of the above mentioned groups brings a distribution of maxima to the line $^1/_2$, $\underline{v}$, 0, in the F^2 series. In the case of group P2/a, in addititlon to a special line, a special plane $\underline{u}$, 0, $\underline{w}$ will arise on which maxima should be distributed.

Table 16 presents such an analysis of all the space groups.

The opposite extreme would be to assume that when a crystal belongs to one of the 192 groups, the appearance of the F^2 series permits one to determine the crystal symmetry uniquely.

There are two limitations:

1) Only the absence and not the presence of a center of inversion can be proved.

2) A true element of symmetry cannot be distinguished from the corresponding pseudoelement of symmetry.

Let us study these statements in some detail. If we know that there is only one atom in a general position in the elementary cell of the crystal, i.e., an atom not on an element of symmetry, then the determination of the space group is really unique. In this case, it is not necessary even to study the F^2 series to find the center of inversion. In fact, the center of inversion doubles the multiplicity of the space group. If we know that there are two atoms in the cell, then the choice between groups Pa and P2/a is solved uniquely to the advantage of the first.

Let us assume, however, continuing the study of the same example, that there are four atoms in the cell. Then either one atom is in a general position and the symmetry of the crystal is P2/a, or two atoms are in general positions and the symmetry of the crystal is Pa.

As has been indicated above, the number of maxima in the case of a centrosymmetric crystal is equal to N single and $(N^2/2) - N$ double, i.e., in our example 4 + 4. In the absence of a center of symmetry all the maxima are single and their number is equal to N (N-1), i.e., twelve.

If, therefore, in studying the F^2 series we find twelve maxima, the absence of a center of inversion is not in doubt. But if the number is eight, then it can only be said that the crystal structure is either centrosymmetric, or only a little different from it. In fact, the F^2 series has a finite resolving power which may be estimated approximately as 0.5 A. More closely placed maxima will merge into one.

Thus, as in other physical methods for determining a center of inversion, the absence but not the presence of a center of inversion may be established by the x-ray method.

Let us now study the second limitation of the x-ray diffraction method in establishing the crystal symmetry.

It has been shown that a symmetry axis is reflected in P-function space by maxima lying on a plane, and a symmetry plane by maxima lying on an axis. However, maxima lying in these planes will also occur in P-function space if there are symmetrically unconnected atoms in the crystal located in the same plane (pseudoaxis of rotation), or in planes shifted by $\frac{1}{2}$ of a lattice constant (pseudoaxis 2_1), etc. The F^2 series will imitate a symmetry plane exactly in the same manner if there are atoms in the crystal distributed along one line (a pseudomirror plane of symmetry) or on two parallel lines shifted by $\frac{1}{2}$ of the lattice translation (pseudoglide plane).

On the basis of what has been said, here, as in the case of a center of inversion, the absence and not the presence of the elements of symmetry can be accurately proved by x-ray diffraction. The absence of maxima along lines or planes of P-function space is a strict proof of the absence of the corresponding elements of symmetry.

The phenomenon of pseudosymmetry is not very widespread. However, one has to keep in mind that the determination of the space group in some cases will be impossible without the application of crystallochemical rules, or without carrying out the structure analysis "to the end," i.e., up to the finding of the atomic coordinates in the unit cell.

5. Picking the Structure Out of the Convolution [45, 46]

When the origin of a crystal is chosen arbitrarily, the structure is characterized by the aggregate of vectors $\mathbf{r}_i$ ($i = 1,\ldots,N$), where N is the number of atoms in the unit cell, and $\mathbf{r}_i$ is a vector connecting the origin of coordinates with the ith atom in the unit cell.

A series of interatomic vectors is characterized by the aggregate of interatomic vectors $\mathbf{r}_{ik} = \mathbf{r}_k - \mathbf{r}_i$. Neglecting the trivial zero solution, the P-function space has $N(N-1)$ maxima at which the vectors $\mathbf{r}_{ik}$ terminate when drawn from the origin of coordinates.

The maxima of the F^2 series are subdivided into N groups related to $N-1$ maxima (not counting the trivial $\mathbf{r}_{11}, \mathbf{r}_{22}$, etc.);

$$\begin{array}{ccccc} * & \mathbf{r}_{12} & \mathbf{r}_{13} & \ldots & \mathbf{r}_{1N}, \\ \mathbf{r}_{21} & * & \mathbf{r}_{23} & \ldots & \mathbf{r}_{2N}, \\ \mathbf{r}_{31} & \mathbf{r}_{32} & * & \ldots & \mathbf{r}_{3N}, \\ \ldots & \ldots & \ldots & \ldots & \ldots \\ \mathbf{r}_{N1} & \mathbf{r}_{N2} & \mathbf{r}_{N3} & \ldots & *. \end{array}$$

It is not hard to see that each group gives the atomic distribution in the crystal with the origin of coordinates at the first, second, third, and so forth atom. In fact

$$\mathbf{r}_{12}=\mathbf{r}_2-\mathbf{r}_1,\quad \mathbf{r}_{13}=\mathbf{r}_3-\mathbf{r}_1,\quad \ldots,\quad \mathbf{r}_{1N}=\mathbf{r}_N-\mathbf{r}_1.$$

Thus, the following fairly obvious theorem is proved: The maxima in P-function space may be subdivided into N groups, each of which gives the atomic distribution in the crystal. These N groups of points are consequently related by translation.

Thus, the problem of determining the atomic configuration in the crystal from the F^2 series would be solved if a method of choosing the maxima that belong to one group, in the sense indicated above, could be obtained.

The subdivision into N groups may take place along the rows or down the columns. The N groups $\mathbf{r}_{1k}, \mathbf{r}_{2k},\ldots$ give groups of points which pass into one another by parallel translations. For the groups $\mathbf{r}_{1k}$ and $\mathbf{r}_{2k}$, the translation vector is equal to $\mathbf{r}_{12} = \mathbf{r}_2 - \mathbf{r}_1$. The same is true also for the groups $\mathbf{r}_{k1}, \mathbf{r}_{k2},\ldots$. Groups of the first type, generally speaking, do not pass into groups of the second type. However they are related. In fact, the groups $\mathbf{r}_{1k}$ and $\mathbf{r}_{k1}$, $\mathbf{r}_{2k}$ and $\mathbf{r}_{k2}$ pass into one another by inversion in the origin of coordinates of P-function space. For brevity, let us call the groups $\mathbf{r}_{1k}, \mathbf{r}_{2k},\ldots$ direct groups and $\mathbf{r}_{k1}, \mathbf{r}_{k2},\ldots$ reciprocal groups.

Thus, there are in all 2N ways of picking out the crystal structure from the maxima of the F^2 series. With a center of symmetry present in the crystal, groups of type $\mathbf{r}_{1k}$ and $\mathbf{r}_{k1}$ are translationally identical, since the operation of a center of symmetry applied to centrosymmetric groups is equivalent to translation.

In the F^2 series there are $N(N-1)$ maxima. The number of all the combinations of $N(N-1)$ taken $N-1$ at a time is equal to $(N-1)!$ Of this number only 2N combinations lead to the result required. What, then, is the method of selecting the combination required, i.e., of finding the atomic groups of the crystal among the maxima in P-function space ?

Let us analyze the group of $N-1$ peaks $\mathbf{r}_{1k}$. Let us call the vector $\mathbf{r}_{12}$ basic. Evidently,

$$\mathbf{r}_{12}+\mathbf{r}_{k1}+\mathbf{r}_{2k}=0 \tag{8}$$

is true for any k, or

$$\mathbf{r}_{12}=\mathbf{r}_{1k}+\mathbf{r}_{k2}.$$

In other words, the basic vector is the diagonal of the parallelogram constructed on two vectors, one of which belongs to the first row (first direct row), and the other to the second column (second reciprocal column). It is evident that the center of the basic vector is a center of symmetry for each pair $\mathbf{r}_{1k}$ and $\mathbf{r}_{k2}$.

Thus the following has been proved: The maxima of P-function space, when inverted in the center of the basic vector, belong only to two groups of points — one direct and one reciprocal. The inversion is performed not only according to the coordinates, but also according to the weights of the maxima; but only when the atomic numbers of the atoms one and two that form the basic vector are equal. In addition, a disturbance of the symmetry not greater than the breadth of the maxima is possible, because of the superposition of the maxima.

Any vector, of course, may be considered as basic. Therefore, the theorem just proved may be formulated in the following manner: If an arbitrary column and row be isolated from the matrix $\mathbf{r}_{ik}$, then the isolated vectors correspond to two groups of points inverted through the center of the vector which is the common element of the row and the column.

For example, the isolated row and column

$$\begin{array}{cc|c|ccc}
* & \mathbf{r}_{12} & \mathbf{r}_{13} & \mathbf{r}_{14} & \mathbf{r}_{15} & \cdots \\
\mathbf{r}_{21} & * & \mathbf{r}_{23} & \mathbf{r}_{24} & \mathbf{r}_{25} & \cdots \\
\mathbf{r}_{31} & \mathbf{r}_{32} & * & \mathbf{r}_{34} & \mathbf{r}_{35} & \cdots \\
\hline
\mathbf{r}_{41} & \mathbf{r}_{42} & \mathbf{r}_{43} & * & \mathbf{r}_{45} & \cdots \\
\hline
\mathbf{r}_{51} & \mathbf{r}_{52} & \mathbf{r}_{53} & \mathbf{r}_{54} & * & \cdots \\
\cdot & \cdot & \cdot & \cdot & \cdot & \cdot
\end{array}$$

are inverted in the center of the vector $\mathbf{r}_{43}$. Moreover, vectors $\mathbf{r}_{4k}$ and $\mathbf{r}_{k3}$, i.e., $\mathbf{r}_{41}$ and $\mathbf{r}_{13}$, $\mathbf{r}_{42}$ and $\mathbf{r}_{23}$, etc., are related by the center of inversion.

The condition just proved limits sharply the number of points from which the crystal structure can be constucted: Instead of $N(N-1)$ maxima, $2(N-1)$ maxima are at hand.

Let us make a further choice of the first atomic configuration for a centrosymmetric crystals. As has been indicated above, N single and $(N^2/2) - N$ double maxima will be present in P-function space. If atoms connected in the crystal by a center of inversion are denoted by the indices k and k', then the maxima $\mathbf{r}_{kk'}$ will be single.

Let us assume that one of the maxima is definitely single. Let us choose the vector of this maximum as basic, and pick out the maxima of the F^2 series inverted in its center. The center of this basic vector is the center of symmetry of a group of atoms (since a vector that passes through the center of symmetry connects only atoms related by this operation). Consequently, the direct group coincides in this case with the reciprocal one, and this choice of maxima leads to a unique solution of the problem.

This is illustrated in Table 17.

TABLE 17

*	r_{12}	r_{13}	$r_{14}\cdots$	$r_{11'}$	$r_{12'}$	$r_{13'}$	$r_{14'}\cdots$
r_{21}	*	r_{23}	$r_{24}\cdots$	$r_{21'}$	$r_{22'}$	$r_{23'}$	$r_{24'}\cdots$
r_{31}	r_{32}	*	$r_{34}\cdots$	$r_{31'}$	$r_{32'}$	$r_{33'}$	$r_{34'}\cdots$
r_{41}	r_{42}	r_{43}	* ...	$r_{41'}$	$r_{42'}$	$r_{43'}$	$r_{44'}\cdots$
. . .	. . .	. . .	. . .	. . . .	. . .	. . .	. . .
$r_{1'1}$	$r_{1'2}$	$r_{1'3}$	$r_{1'4}\cdots$	*	$r_{1'2'}$	$r_{1'3'}$	$r_{1'4'}\cdots$
$r_{2'1}$	$r_{2'2}$	$r_{2'3}$	$r_{2'4}\cdots$	$r_{2'1'}$	*	$r_{2'3'}$	$r_{2'4'}\cdots$
$r_{3'1}$	$r_{3'2}$	$r_{3'3}$	$r_{3'4}\cdots$	$r_{3'1'}$	$r_{3'2'}$	*	$r_{3'4'}\cdots$
$r_{4'1}$	$r_{4'2}$	$r_{4'3}$	$r_{4'4}\cdots$	$r_{4'1'}$	$r_{4'2'}$	$r_{4'3'}$	* ...

The single vector $\mathbf{r}_{11'}$ is considered as basic. All the maxima of the isolated row and column, except $\mathbf{r}_{11'}$, merge in pairs and become double. The maxima $\mathbf{r}_{21'}$, and $\mathbf{r}_{22'}$, $\mathbf{r}_{31'}$ and $\mathbf{r}_{13'}$, etc., coincide, since the vectors $\mathbf{r}_{1'} - \mathbf{r}_2$ and $\mathbf{r}_{2'} - \mathbf{r}_1$, $\mathbf{r}_{1'} - \mathbf{r}_3$ and $\mathbf{r}_{3'} - \mathbf{r}_1$ are parallel. The maxima $\mathbf{r}_{1k}$ and $\mathbf{r}_{k1'}$, $\mathbf{r}_{12'}$ and $\mathbf{r}_{2'1'}$, $\mathbf{r}_{13'}$ and $\mathbf{r}_{3'1'}$, etc., are inverted through the center of the vector $\mathbf{r}_{11'}$.

Let us assume now that the single maxima are poorly resolved and that the basic vector has been drawn into a double maximum. Then the two groups of points $\mathbf{r}_{1k}$ and $\mathbf{r}_{k2}$ will be picked out by an inversion in the center of the basic vector; among them, along with the double maxima, there will be single ones. Since the crystal has a center of symmetry, both groups of points are related by a translation. Thus, after choosing, we have two "crystal structures" superimposed on one another with a parallel translation. How can these groups of points be separated from one another?

The most direct method is a search for the translation vector of the two "structures." Evidently, the whole pattern may be subdivided into N−1 pairs of maxima, connected by the vector ρ, and, consequently, the group $\mathbf{r}_{1k}$ will be automatically separated from the group $\mathbf{r}_{k2}$.

Let us pass over to the discussion of the case of a noncentrosymmetric crystal. According to the former scheme of inverting in the center of the basic vector $\mathbf{r}_{12}$, we select two groups $\mathbf{r}_{1k}$ and $\mathbf{r}_{k2}$. These two groups are not parallel to each other. To isolate the points of one group we proceed in the following manner.

We select arbitrarily three points of the first group. These will be: The origin of P-function space, the end of the basic vector $\mathbf{r}_{12}$ and any one of the selected maxima which we shall call $\mathbf{r}_{13}$. Let us now study the two points A and B inverted through the center of $\mathbf{r}_{12}$; let these be the ends of the vectors $\mathbf{r}_{14}$ and $\mathbf{r}_{42}$. However, we don't know which of these vectors corresponds to point A or B. Vector $\mathbf{r}_{14}$ and not vector $\mathbf{r}_{42}$ enters into the group which we have begun to constuct. Let us connect point A with the end of vector $\mathbf{r}_{13}$. If A is $\mathbf{r}_{14}$, then this line segment is equal to $\mathbf{r}_{14} - \mathbf{r}_{13} = \mathbf{r}_{43}$; this is an interatomic vector and, consequently, there is a vector from the origin of coordinates equal and parallel to the segment drawn from A to the end of $\mathbf{r}_{13}$. If A is $\mathbf{r}_{42}$, then the segment is equal to $\mathbf{r}_{42} - \mathbf{r}_{13} = \mathbf{r}_2 - \mathbf{r}_4 - \mathbf{r}_1 + \mathbf{r}_3$; this is not an interatomic vector and in this case there is no vector from the origin of coordinates equal and parallel to the one drawn.

By the method just indicated, we may experiment with all the pairs of maxima $\mathbf{r}_{1k}$ and $\mathbf{r}_{k2}$, and select the maxima $\mathbf{r}_{1k}$ from among them.

What difficulties may arise in putting this method into practice? Evidently, this method is doomed to be unsuccessful if a large number of maxima are not resolved. Projections of the F^2 series may therefore be used with this method only in the simplest cases. On the other hand, the possibilities of three-dimensional series are rather great, as follows from a preliminary calculation. In the case of four molecules of eight atoms in the unit cell the general number of peaks of the F^2 series is equal to almost 1000. In the usual division of the cell the calculation of the series is performed at approximately $50 \times 50 \times 50 = 125{,}000$ points. Thus, on the average, there is one peak to 100 points. When there are 32 atoms in the cell its volume will be of the order of 1000 A^3; consequently, the volume per peak will be of the order of 1 A^3. When the resolving power of the series is of the order 0.5 A there will be in this volume, not one but 25 resolved maxima. Thus, the merging of maxima due to their accidental proximity will be relatively rare, and the method just developed has a chance of success in the majority of extremely complicated cases.

Until recently, structure analyses by the F^2-series method were carried out only for structures with one or two heavy atoms.

Let us assume, for example, that there are two heavy atoms in the cell and a number of light ones. Then $\mathbf{r}_{12}$ gives one strong maximum in P-function space. The column and row of the matrix $\mathbf{r}_{ik}$ which cross at $\mathbf{r}_{12}$ will give maxima of intermediate strength; and all the remaining $\mathbf{r}_{ik}$ will give weak maxima. Thus, the heavy atom automatically performs that selection into groups the artificial use of which has been discussed above. It is in just such a case that V. V. Sanadze and G. S. Zhdanov [47] established for the first time the relationship of the points of the F^2 series to a center of inversion in the center of an interatomic vector.

Of great interest, however, is the use of the connection between the F^2 series and the crystal structure for crystals made up of atoms with almost the same scattering power. There have not been sufficient experimental analyses by this method up to the present time.

The methodology just described is of great interest for structures without heavy atoms. A structure without a center of symmetry with atoms of the same scattering power is especially interesting for checking the possibilities of direct structure determination from three-dimensional series of interatomic vectors.

We attach great importance to the use of this methodology together with the methods of direct sign determination. Both methods lead to exceptional simplification of the crystal structure analysis. The most cumbersome stage of the analysis – the calculation of structure amplitudes – becomes superfluous.

The above discussions show that there is a way, not only in principle, but also in practice, of finding the structure directly from the F^2 series.

It should not be forgotten that symmetry provides a number of important simplifications. We have already stressed that for a lattice with a center of symmetry the F^2 series contains N single maxima (all the remaining ones are double) that give the structure directly. If all these maxima can be clearly identified, then the finding of the structure is trivially simple. The symmetry axes and planes place the maxima in special positions, etc.

The methodology that has been analyzed here should lead to a positive result in all cases. The possibility for repeated checking within the framework of the method itself is very important (any vector may be taken as basic). We shall not discuss here the use of F^2 series in structure analyses when there is a small number of heavy atoms in the cell, since these methods are well known and are described in monographs. For the same reason, we do not discuss the methods of computing Fourier series here.

6. The Difference Inversions

It has been shown that, in principle, it is possible to isolate from the Patterson function a system of interatomic functions connected by inversion through the center of the "basic" vector. Depending on the selection of this vector, the system isolated will represent the crystal structure or two structures superimposed on one another.

The overlapping of interatomic functions may substantially complicate this work. The following auxiliary method proves to be useful. Let us isolate a cell constructed on the centers of the basic vectors, and perform the following operation with each value of $P(\mathbf{r})$ within this cell. Let us study a point $\mathbf{r}$ within the isolated cell, and the point $\mathbf{r}'$ outside it connected with $\mathbf{r}$ by inversion in the pseudocenter.

$P(\mathbf{r})$ and $P(\mathbf{r}')$ are different. However, the crystal structure picked out from $P(\mathbf{r})$ and $P(\mathbf{r}')$ must be the same for the ideal case of nonoverlapping interatomic functions. Consequently, we shall approach more closely to the result sought if the following operation is performed for all the points of the isolated cell: Let us preserve the value $P(\mathbf{r})$ if $P(\mathbf{r}) < P(\mathbf{r}')$, and substitute the value $P(\mathbf{r}')$ for $P(\mathbf{r})$ if $P(\mathbf{r}) > P(\mathbf{r}')$.

Let us denote the new function $P_1(\mathbf{r})$ and extend it anew over the whole of space. The pseudocenter of inversion will now become a true center of inversion for $P(\mathbf{r})$.

If the interatomic functions did not overlap, then $P_1(\mathbf{r})$ would represent a structure (or two superimposed structures) of atoms having the shape of interatomic functions.

The transition from $P(\mathbf{r})$ to $P_1(\mathbf{r})$ makes sense, because during this transition, maxima in interatomic functions may appear that are discernible on $P(\mathbf{r})$ as points of inflection only.

Because of the overlapping of interatomic functions $P_1(\mathbf{r})$ will not quite resemble the structure. It is possible to attempt to approach a structure by repeating the operation, i.e., to extend $P_1(\mathbf{r})$ to infinity and to perform the inversion in the center of another basic vector of the original $P(\mathbf{r})$ function. The practical usefulness of such repeated minimalizations has not yet been shown. The theoretical bases of repeated minimalizations are in a rather confused state.

Buerger [48] has suggested another technique for performing the minimalization method. The essence of his method is quite similar to the one discussed above. See also B. K. Vainshtein's article [49].

7. Sharpening of Convolutions

The interatomic function, as has been shown in Section 2, is rather blurred. This, together with the large number of possible peaks in any even slightly complicated crystal, leads to the frequent superposition of peaks. Patterson, even in his first paper of 1935, suggested a method of sharpening the F^2 series, which amounts to the substitution of point atoms for real ones.

This sharpened series has the form

$$\hat{P}(\mathbf{r}) = \sum \hat{F}^2 \cos 2\pi \mathbf{H}\mathbf{r},$$

where $\hat{F}$ is a unitary structure amplitude.

Clearly, $\hat{P}(\mathbf{r})$ is a divergent sum.

Nevertheless, it is interesting to note that in a number of cases the influence of the fluctuations due to cutoff on the form of $\hat{P}(\mathbf{r})$ is negligible. If, on the other hand, the researcher considers the termination of the series to be important, then, generally speaking, instead of interatomic point functions, functions of any form may be introduced. For this purpose, it is sufficient to multiply $\hat{F}$ by some decreasing function of H. The most convenient method of introducing this conventional factor is in the form of $e^{-\alpha|H|}$ or $e^{-\alpha H^2}$ arbitrarily selecting α.

There are indications of the considerable usefulness of applying fully or partially sharpened F^2 series when interpreting fairly complex structures. However, it would be interesting to study F^2 series with the aid of a computing machine in order to establish methods of selecting a modifying function which gives a sufficient convergence and guarantees the elimination of false peaks during sharpening.

It is useful to accompany the sharpening of an F^2 series with the removal of the zero peak which, as is obvious from Section 1, does not furnish any data concerning interatomic distances and, at the same time, may conceal peaks close to the origin of coordinates. Since the zero peak is the Fourier transform of Σf_j^2 (see page 209), its removal boils down to the substitution for F^2 or $F^2 - \Sigma f_j^2$. But if the transition to unitary structure amplitudes has been made, then $\hat{F}^2$ should be substituted by

$$\hat{F}^2 - \sum n_j^2.$$

Thus, the formula for a sharpened F^2 series with the zero peak removed has the form

$$\hat{P}(\mathbf{r}) = \sum e^{-\alpha |H|} (\hat{F}^2 - \sigma^2) \cos 2\pi \mathbf{H}\mathbf{r}. \tag{9}$$

The unitary amplitudes are found by means of the procedure stated in chapter III. The coefficient α is selected by trial and error.

CHAPTER VI

METHODS OF OBTAINING AGREEMENT BETWEEN THE MEASURED AND CALCULATED STRUCTURE AMPLITUDES

1. Antagonistic Reflections

X-ray structure analysis does not determine directly the mean electron density of the crystal. Experiment gives only the moduli of the structure amplitudes. To construct an electron density series we need information about the phases (signs) of the amplitudes which is obtainable by the direct methods discussed in chapters III and IV, or by calculating structure amplitudes, if at least approximate atomic coordinates are known.

In both cases the calculation of the signs of amplitudes becomes simple only with the simplifying assumption that the electron density of the crystal may be expressed as a sum of spherical atomic densities.

Deviations of atoms from spherical symmetry due to the deviation of the electron distribution of the atom from this symmetry, especially because of thermal vibrations, may be quite significant and lead to the following effect.

Let $\mathbf{r}_j$ be the "true" atomic centers, i.e., the maxima of the correctly measured electron density.

The measured (with the subscript o) structure amplitudes

$$F_o(\mathbf{H}) = \sum_{j=1}^{N} f_j(\mathbf{H}) \exp 2\pi i \mathbf{H}\mathbf{r}_j \quad (1)$$

may be expressed by the vectors $\mathbf{r}_j$ and by the atomic factors. Moreover, the $f_j(\mathbf{H})$ depend not only on the magnitude, but also on the direction of the vector $\mathbf{H}$.

In the initial stage of the analysis this dependence cannot be known, since it is determined by the structure. One must use the atomic factor averaged over all directions:

$$\bar{f}(\mathbf{H}) = \frac{1}{4\pi H^2} \int_{S_H} f(\mathbf{H})\, dS_H, \quad (2)$$

where S_H is a spherical surface of radius H in reciprocal space. Consequently, the calculated (subscript c) structure amplitudes may be expressed by the formula

$$F_c = \sum_{j=1}^{N} \bar{f}_j(H) \exp 2\pi i \mathbf{H}\mathbf{r}_j. \tag{3}$$

It is clear that in the case of a crystal without a center of inversion, F_0 and F_c will have, generally speaking, different phases. However, such an effect, as we shall now see, is also quite possible for centrosymmetric crystals in the case of weak reflections. If F_0 and F_c have different signs for "true" $\mathbf{r}_j$ values, then such reflections will be called antagonistic.

Evidently, the finding of the signs of amplitudes by direct methods is possible only for nonantagonistic reflections. The finding of the signs of these reflections is possible in principle only in the last stage of the analysis when the structure is already known with a high degree of accuracy. In this case one might attempt to find f (**H**) and, consequently, discover whether the sign of F_0 in (1) coincides with that of F_c, calculated by (3) for spherical atoms.

Let us first try to establish differences between F_0 and F_c which arise because the anisotropy of the atomic electron cloud has not been taken into account, and let us find the conditions under which reflections must be associated with antagonistic ones.

The expression for the structure amplitude (1) may be written in the form (see chapter II, section 4)

$$F_o(\mathbf{H}) = \sum_{j=1}^{N} f_{0j}(\mathbf{H})\, e^{-B_j H^2} \exp 2\pi i\, \mathbf{H}\mathbf{r}, \tag{4}$$

where $B_j = 2\pi^2 u_j^2(\mathbf{H})$, and $u_j^2(\mathbf{H})$ is the mean square displacement of the jth atom in the **H** direction.

As has already been said, in the initial stage of a structure analysis the dependencies of the atomic factor f_0 and of the average displacement of the atom u on the direction of the vector **H** are unknown to us, since a knowledge of the structure is required for this. Moreover, at this stage neither is there any information on the average values of f_0 and u taken over all directions of the vector **H**. Nevertheless, let us assume that it has been possible to carry out this work by selection (trial and error method), and that for each atom a mean isotropic factor $\bar{f}_j(H)$ has been found, connected with the true one by (2).

Let us compare the observed and the calculated values of the structure amplitudes:

$$F_c = \sum_{j=1}^{N} f_j \exp 2\pi i \mathbf{H}\mathbf{r}_j;$$

$$F_o = \sum_{j=1}^{N} f_{0j} e^{-B_j H^2} \exp 2\pi i \mathbf{H}\mathbf{r}_j.$$

Let us find the mean square divergence between these magnitudes $\overline{(\Delta F)^2} = \overline{(F_o - F_c)^2}$. We obtain:

$$\Delta F = \sum_{j=1}^{N} \{f_{0j} e^{-B_j H^2} - \bar{f}_j\} \exp 2\pi i\, \mathbf{H}\mathbf{r}_j.$$

whence

$$\overline{(\Delta F)^2} = \sum_{j=1}^{N} \overline{\{f_{0j} e^{-B_j H^2} - \bar{f}_j\}^2} = \sum_{j=1}^{N} \overline{(\Delta f_j)^2}. \tag{5}$$

But

$$\bar{f}_j = \overline{f_{0j} e^{-B_j H^2}} = \bar{f}_{0j} \overline{e^{-B_j H^2}},$$

since the anisotropies of $\bar{f}_{0j}$ and $\overline{u^2}$ are independent of one another.

Since $B_j H^2$ is a quantity that varies within narrow limits for a given H, then $\overline{e^{-B_j H^2}} = e^{-\bar{B}_j H^2}$ approximately. Thus

$$\bar{f}_j = \bar{f}_{0j} e^{-\bar{B}_j H^2}. \tag{6}$$

Let f_{0j} differ from $\bar{f}_{0j}$ by Δf_{0j}, and B_j from $\bar{B}_j$ by ΔB_j. Then the deviation of the atomic factor from the average may be expressed as

$$\Delta f_j = e^{-B_j H^2} \Delta f_{0j} - \bar{f}_{0j} e^{-B_j H^2} H^2 \Delta B_j,$$

or

$$\frac{\Delta f_j}{f_j} = \frac{\Delta f_{0j}}{f_{0j}} - H^2 \Delta B_j. \tag{7}$$

The mean square deviation will obviously be equal to

$$\overline{\left(\frac{\Delta f_j}{f_j}\right)^2} = \overline{\left(\frac{\Delta f_{0j}}{f_{0j}}\right)^2} + H^4\ \overline{(\Delta B_j)^2}. \tag{8}$$

Let us recall that the averaging must be performed over all directions of **H** for the absolute value $|\mathbf{H}|$ under study.

McWeeny's [6] work enables one to determine the first term of the sum fairly well. As has been shown by this author (see chapter I), the atomic factor may be calculated by the formula

$$f_0 = f_0^{\parallel} \cos^2\varphi + f_0^{\perp} \sin^2\varphi,$$

where the angle φ is laid off from the symmetry axis of the electron cloud (this formula is not necessary for electrons of spherical atoms, for example, lithium, beryllium). The magnitudes of $f_0^{\parallel}$ and $f_0^{\perp}$ have been calculated for light atoms $\overline{(\Delta f_0)^2}$ is calculated from the formula

$$\overline{(\Delta f_0)^2} = \frac{1}{2}\int_0^{\pi} (f_0^{\parallel}\cos^2\varphi + f_0^{\perp}\sin^2\varphi - \bar{f}_0)^2 \sin\varphi\, d\varphi, \tag{9}$$

where

$$\bar{f}_0 = \frac{1}{3} f_0^{\parallel} + \frac{2}{3} f_0^{\perp}.$$

Integration gives

$$\overline{(\Delta f_0)^2} = \frac{4}{45}(f_0^{\parallel} - f_0^{\perp})^2. \tag{10}$$

whence

$$\sqrt{\frac{\overline{(\Delta f_0)^2}}{f_0^2}} = 0.9\,\frac{1 - \dfrac{f_0^{\perp}}{f_0^{\parallel}}}{1 + 2\dfrac{f_0^{\perp}}{f_0^{\parallel}}}. \tag{11}$$

McWeeny's figures show that this quantity does not exceed 10-12%. Maximum values occur at average scattering angles (see Table 1 in chapter II).

We have no reliable data at our disposal for the evaluation of the second term of (8). Experiment shows that for atoms of organic substances $\overline{B}$ lies within the limits 0.5-1 A^2. For other crystals the lower boundary may be further decreased twofold. A number of structure analyses can be cited in which the magnitudes of B for different directions differ several-fold, even approaching zero in certain directions (see, for example, Zener's data for zinc, for FH_2 [51] and for methanol [50].

Let us take an ellipsoid of revolution for the mean square atomic displacement. Then

$$\bar{B}_j = B_j^{\parallel} \cos^2 \varphi + B_j^{\perp} \sin^2 \varphi.$$

Carrying out the same calculations as for f we obtain, analogously to (10) and (9)

$$\sqrt{\overline{|\Delta B|^2}} = \frac{2}{\sqrt{45}} (B_j^{\parallel} - B_j^{\perp}) \text{ and } \bar{B}_j = \frac{1}{3} B_j^{\parallel} + \frac{2}{3} B_j^{\perp}.$$

The anisotropy will be the most acute when $B_j^{\perp} = 0$. In this case $(\overline{|\Delta B_{max}|}) = 0.9\bar{B}$.

Thus, it is improbable that we shall make a substantial error if we assume

$$\sqrt{\overline{|\Delta B|^2}} = 0.45B. \tag{12}$$

Since $\bar{B}_j$ is a quantity that is usually within the limits 0.5 – 1 A^2 (nearer to 1), then for large scattering angles the first term in (7) will be smaller than the second. The anisotropy of the molecular vibrations shows itself more strongly in the scatter of the values of the atomic factor, than does the anisotropy of the electron cloud of the atom at rest.

Thus, starting from (8) and (12) we obtain

$$\sqrt{\overline{\left(\frac{\Delta f}{f}\right)^2}} \approx 0.45 H^2 \bar{B}_j. \tag{13}$$

For $B_j = 1$ A^2, and taking into account that $e^{-B_j H^2} = e^{-4\left(\frac{\sin\vartheta}{\lambda}\right)^2}$, we obtain magnitudes such as, for example, for CuK_α radiation

ϑ, degrees	30	60	90
Average deviation, %	18	60	75

Let us make a definite underestimate of this deviation, that is, let us assume $\overline{(\Delta f_j/f_j)^2} = 0.1$.

Evaluating the mean square deviation of the structure amplitudes (5), we obtain

$$\overline{(\Delta F)^2} \approx 0.1 \sum \overline{f_j^2}. \quad (14)$$

Since (see chapter III) $\overline{F^2} = \Sigma f_j^2$, then $(\Delta F)^2 / F^2 = 0.1$, which agrees well with the usual evaluations of this quantity from concrete examples.

Let us now evaluate the mean square deviation of a unitary structure amplitude. Substituting $F = Z\hat{f}\hat{F}$ and $f = Z\hat{f}$ in (14), we obtain

$$\overline{(\Delta \hat{F})^2} \approx 0.1 \sum n_j^2.$$

i.e.,

$$\sqrt{\overline{(\Delta \hat{F})^2}} = 0.35 \sqrt{\sum n_j^2} = 0.35\sigma, \quad (15)$$

and for identical atoms,

$$\sqrt{\overline{(\Delta \hat{F})^2}} = 0.35 \frac{1}{\sqrt{N}}.$$

Thus, the values of the unitary structure amplitudes (which are calculated on the assumption that the atoms are spherical) may be evaluated with an accuracy not exceeding ± 0.1 for a ten-atom structure, and not exceeding ± 0.035 for a 100-atom structure.

We see that in the initial stage of a structure analysis unitary amplitudes smaller than 0.2-0.3 σ are not useful. Such amplitudes may have different signs depending on the nature of the thermal vibrations (or other anisotropy) in comparison with the signs of amplitude calculated for spherically symmetrical atoms with an isotropic atomic factor f_j. These are the antagonistic amplitudes.

It is not hard to see that our estimation of the amplitudes which are not useful coincides with experiment. Since the magnitudes of $\hat{F}$ are distributed according to the Gaussian law, the number of amplitudes smaller than 0.25 σ is 20%. Actually, as a rule, in carrying out a structure analysis we use about 80% of the reflections that are included in the limiting sphere.

The determination of the signs of amplitude for which $\hat{F}<0.25$ is a very complicated problem. The direct methods discussed in chapters III and IV are not applicable here, since they are based on the assumption of the spherical symmetry of atoms. The direct calculation of the magnitudes of $\hat{F}$ from the atomic coordinates for strong reflections is likewise incorrect if there is no information about the anisotropy of $\underline{f}$.

It is quite untenable to introduce these reflections without taking into account the anisotropy in the process of minimalizing the difference between F_c and F_o.

In principle, attempts may be made to take into account the anisotropy of $\underline{f}$ and, consequently, of decreasing $(\overline{\Delta F})^2$. Using the values of nonantagonistic amplitudes, it is possible to construct the so-called electron density difference series; i.e., the series

$$\Delta(xyz) = \frac{1}{V} \sum (F_o - F_c) \exp - 2\pi i \mathrm{H}_{hkl}\mathbf{r}. \tag{16}$$

The series $\Delta(xyz)$ shows the deviation of the atomic shape from spherical. Since the atomic shape is the Fourier transform of the atomic factor, attempts to take into account the anisotropy followed by sign determinations of weak reflections may be made which may then be introduced into the electron density series. Such work has been carried out only in isolated instances.

2. The R-Factor and the Correlation Coefficient

The comparison of the structure amplitudes (F_c) calculated from the model with the corresponding amplitudes (F_o) measured experimentally is often used as a criterion for the accuracy of the model.

Since both the structure amplitudes F_o and F_c are subject to the Gaussian distribution, the well known probability theory of normal correlation may be used for a comparison of these magnitudes.

The correlation between F_o and F_c can be analyzed in the greatest detail by the construction of the function $P(F_o, F_c)$, where $P\Delta F_o \Delta F_c$ is the number of reflections for which the F_o and F_c values lie respectively between $F_o + \Delta F_o$ and $F_c + \Delta F_c$. This number, according to the theory, must be close to the following expression (see [18], page 364, formula 94)

$$\frac{1}{2\pi\sigma^2\sqrt{1-k^2}} \int\limits_{F_o}^{F_o+\Delta F} \int\limits_{F_c}^{F_c+\Delta F_c} \exp - \frac{1}{2(1-k^2)} \left[\frac{x^2+y^2-2Rxy}{\sigma^2}\right] dx\, dy.$$

Here $\sigma^2 = \overline{F^2}$. We consider σ to be equal for both the calculated and and the experimental amplitudes, and k is the correlation coefficient equal to $k = \overline{F_o F_c} / \sigma^2$.

The function $P(F_o, F_c)$ is a complete criterion for the nature of the correlation between the measured and the calculated amplitudes. If this function were close enough to the double integral just written (and this will occur if the amplitudes, as sums of a large number of terms, satisfy the conditions of Liapunov's theorem, which in practice is realized for a crystal which has more than 20-40 atoms in its cell), then the correlation coefficient would characterize the convergence of F_o and F_c quite uniquely.

The correlation coefficient k is equal to unity when F_o and F_c coincide completely, to zero when there is no correlation, and lies between these limits when the correlation is incomplete.

It is easy to calculate the correlation coefficient; its mathematical meaning is quite clear. Its application can be recommended as a criterion for the degree of reliability of the structure.

Let us write

$$G^2 = \frac{\Sigma (F_o - F_c)^2}{\Sigma F^2} = \frac{\overline{(F_o - F_c)^2}}{\overline{F_o^2}}.$$

It is not hard to see that the G-factor is uniquely connected with k

$$\overline{(F_o - F_c)^2} = \overline{F_o^2} - 2\overline{F_o F_c} + \overline{F_c^2} = 2\overline{F^2} - 2\overline{F_o F_c}.$$

Dividing by $\overline{F_o^2}$, we obtain

$$\frac{\overline{(F_o - F_c)^2}}{\overline{F_o^2}} = 2 - 2k,$$

i.e.,

$$G = \sqrt{2 - 2k}.$$

When there is no correlation, $G = \sqrt{2}$.

In the practice of structure analysis the criterion used for the convergence of F_o and F_c is the so-called reliability index:

$$R = \frac{\Sigma \left| |F_o| - |F_c| \right|}{\Sigma |F_o|} = \frac{\overline{\left| |F_o| - |F_c| \right|}}{|\overline{F_o}|},$$

which has no obvious or clear theoretical meaning. The defect of the R-factor is its dependence on the symmetry of the crystal. This is due to the fact that $|\overline{F}|$ is dependent on symmetry, whereas $\hat{F}^2$ is not (see chapter III).

For $\underline{k}$ close to unity, the G and R factors are rather close to one another. From then on G and R begin to diverge and G becomes greater than R. This follows from Wilson's [52] calculation, which shows the value of R when there is no correlation equal to 0.59 and 0.83 for crystals with and without a center of symmetry respectively. As has been indicated above, G is equal to $\sqrt{2}$ under these conditions.

The values of the reliability factors for certain structure determinations with accuracies typical of those that have been obtained are given in Table 18.

Two numbers are given in each square of Table 18. Their origin is the following. Weak reflections are given in the table for which only the upper boundary can be stated: $F_o < a$. These reflections are often neglected in calculating the reliability indices. It is clear that the impression of correlation is more favorable (the lower figure). If in the calculation $F_{meas} = a$ is assumed in computing the reliability indices, then the upper figure results.

A more rigorous procedure has been suggested by Hamilton [53]. Accepting the Gaussian distribution for structure amplitudes, it is not hard to show that the mean value of F^2 for the unobserved reflection for the absence and presence of an inversion center equals $a^2/2$ and $a^2/3$, respectively.

Returning to Table 18, we see that the values R and G are very close to one another. Nevertheless, we recommend the use of the coefficient $\underline{k}$ or the G-factor, since it alone is directly connected to the correlation coefficient $\underline{k}$.

The last two columns of Table 18 show that the values of the relative mean deviation between the measured and calculated amplitudes there given are substantially larger than the G-factor. We deemed it necessary to give these figures, because sometimes it is erroneously assumed that the R-factor characterizes directly the error of measurement and the other divergencies between F_o and F_c.

TABLE 18

	R	G	k	$\frac{\overline{\lvert F_0\rvert - \lvert F_c\rvert}}{\lvert F_0\rvert}$	$\sqrt{\frac{\overline{(F_0 - F_c)^2}}{F_0^2}}$
Naphthalene	0.202 0.184	0.175 0.168	0.985 0.986	0.33 0.29	0.45 0.41
Anthracene	0.221 0.178	0.194 0.179	0.981 0.984	0.39 0.26	0.60 0.38
Hexamethyl-enedianune dihydrochlor-ide	0.174 0.141	0.156 0.139	0.988 0.990	0.30 0.21	0.45 0.30

3. General Features of Approximation Methods

Two definitions of an accurate crystal structure may be given. First as the aggregate of the maxima of a convergent electron density series:

$$\rho(\mathbf{r}) = \frac{1}{V} \sum_{\mathbf{H}} F_o(\mathbf{H}) \exp - 2\pi i \mathbf{H}\mathbf{r}, \tag{17}$$

in which the signs of all the coefficients are accurately determined. Second as an aggregate of vector coordinates for which

$$\sum (F_c - F_o)^2 \tag{18}$$

is a minimum. For an estimate of the minimum of this sum, a knowledge of $f(\mathbf{H})$ and not of $f(H)$ is necessary for the calculation of F_C.

Both definitions are of a theoretical nature. Actually, in the electron density series the small F's are absent, either because they were not measured or, being antagonistic, their signs could not be determined. Formula (18) enables us to find the coordinates of the centers and the shapes of the atoms which are the best for a limited experiment.

It is clear that neither of these problems can be solved rigorously. The electron density may be calculated with an accuracy limited by the absence of data for weak reflections. The mean square deviation between the measured amplitudes and the calculated ones may be used for finding a structure which best approximates the true one under the conditions of a necessarily limited experiment. Thus, the structures found under the real conditions (17) and (18) may not coincide.

We have mentioned in the preceding paragraph that the shapes of the atoms are determined from the weak reflections. Therefore, the deviations between the structures determined from (17) and (18) never exceed 0.02-0.04 A, and these magnitudes are close to the experimental errors.*

The definition (18) is widely used by many English researchers, i.e., a correct structure is one for which F_0 converges the best with F_c (yet in an overwhelming number of cases, an atomic shape is selected which has no objective physical meaning).

From our point of view, the only rigid definition of a structure is given by (17). The inevitable inaccuracy of (17) only emphasizes the fact that a structure, like every other physical magnitude, may be found with limited accuracy only. It is this last circumstance which is lost in applying (18). With the aid of (18) it is possible "to improve" a structure, shifting the atomic centers by 0.01, 0.001, and even by 0.0001 and arbitrarily changing the shapes of the atoms. Such a procedure is devoid of physical meaning.

Does it follow from this that the correlation between F_0 and F_c should not be used in structure analysis ? Definitely not. There are problems in which the use of (18) is expedient. Let us analyze a number of cases of structure analysis.

a) A fraction of the signs of the nonantagonistic amplitude has been determined; a part of the signs may be incorrect.

b) A fraction of the signs of the nonantagonistic amplitudes has been determined; all the signs are correct.

c) The signs of all the nonantagonistic amplitudes have been determined; a part of them may have been determined incorrectly.

d) The signs of all the nonantagonistic amplitudes have been determined correctly.

Let us call the structure analyses limited by points "a" and "b" incomplete; if points "c" and "d" are added, then we shall speak of a complete analysis. If, finally, attempts have been made to find the signs of the antagonistic reflections and to include them in the series, then this analysis will be termed exhaustive. Now we can elucidate the place of methods of approximation at each stage of the analysis.

---

* Translator's note: Modern crystal structure analysis can be at least five times more accurate than this.

At stage "a" it is expedient to use approximation methods. Signs that have been incorrectly selected are revealed.

At stage "b" these methods are expedient, because an electron density series (17) will give a greater error if some of the strong reflections have not been included.

At stage "c" these methods are expedient, since with their aid, incorrect signs are found.

At stage "d" these methods are not expedient. An electron density series will give an objective result at this stage, and by approximating F_0 to F_c an atomic structure will be found as close as possible in shape to the one under study. The rounding out of a complete analysis is, from our point of view, the construction of $\rho(\mathbf{r})$ [4].

In an exhaustive analysis, approximation methods are almost impossible in practice, since at this stage one has to give up the concept of spherical atoms.

There are the following methods of approximating F_0 and F_c.

1) The trial method. F_c is calculated for different models and compared with the result obtained from experiment. The best model is the one with a minimum R.

2) The method of constructing a set of electron density series. Atomic coordinates are found from the first series. From these data F_c' is calculated. The F_c' signs are applied to F_0. A second electron density series is constructed. Again the $\mathbf{r}_j$ are found from the maxima of the series and a second F_c'' series is calculated. If the signs of a part of F_c'' differ from those of F_c', the corresponding corrections are introduced into the series and the series calculated again.

The calculation is terminated when the F_c'' signs coincide with those of F_c'. If antagonistic reflections have been included in the series, then "there will be no end" the signs of all these F_c'' will always be opposite to the F_c' signs. These reflections must be excluded from the series.

3) A certain variation of this method is the same, except that the repeated calculation is made not of the series $\rho(\mathbf{r})$, but of its derivative. Let us call it the differential method of structure refinement.

4) The method of minimizing

$$\sum (F_o - F_c)^2$$

or an analogous expression: a) the method of least squares (Hughes), b) the method of "the steepest descent" (Booth).

5) The method of difference series. This consists in studying the series

$$\rho_o - \rho_c = \sum (F_o - F_c) \exp - 2\pi i \mathbf{Hr}.$$

The distribution of the maxima of these series shows the direction in which the coordinates of the atomic centers should be shifted so as to approximate F_o by F_c.

These methods will be studied in the following paragraphs.

4. Booth's Method of "Steepest Descent"

The approximating of F_o and F_c can be accomplished by using the amplitudes themselves, their squares; the unitary amplitudes, or their squares; it is also possible to use other functions of the structural parameters as well. Under parameters may be understood the atomic coordinates, Euler's molecular angles, or other magnitudes, through which structure amplitudes may, in principle, be expressed. Assume that the structure has n parameters x_i. Let us denote by Φ_o the measured functions, and by Φ_c the calculated functions of hkl and the atomic coordinate.

Since different amplitudes may play a different part in the accuracy of determining the coordinates, the "weight" of their squared difference $(\Phi_o - \Phi_c)^2$ must be introduced and we denote it by w^2. Then the problem is to find the minimum of the quantity

$$R = \sum_{hkl} w^2 (\Phi_o - \Phi_c)^2. \tag{19}$$

R is a function of the n variable magnitudes x_i. The structural parameters for which R is of some significance, define a point lying on the $(n-1)$-dimensional surface $R = \text{const}$.

Let us imagine that with the aid of (19) we have constructed a series of such surfaces for various R. Then the problem of minimalization may be formulated as the problem of seeking the shortest descent from a surface with a larger R to one with smaller R as close to zero as possible.

The direction of the shortest descent is the line grad R, which by definition, is perpendicular to all the surfaces of equal R that it intersects. If grad R were known then the problem could be solved directly.

In dealing with the minimalization we know the position of only the one original R surface, and therefore, we know only the initial direction of the line of descent.

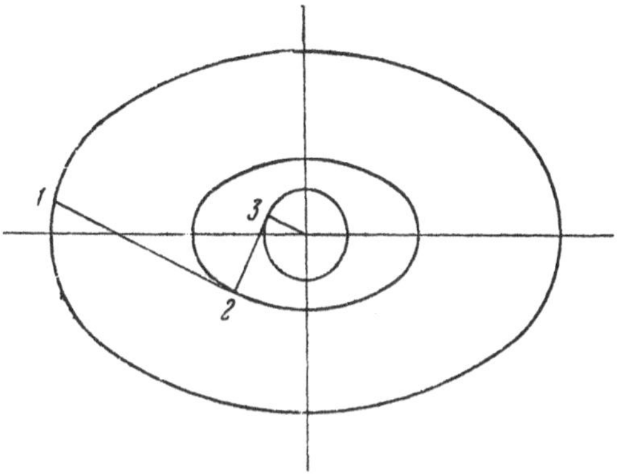

Fig. 21. The method of "steepest descent."

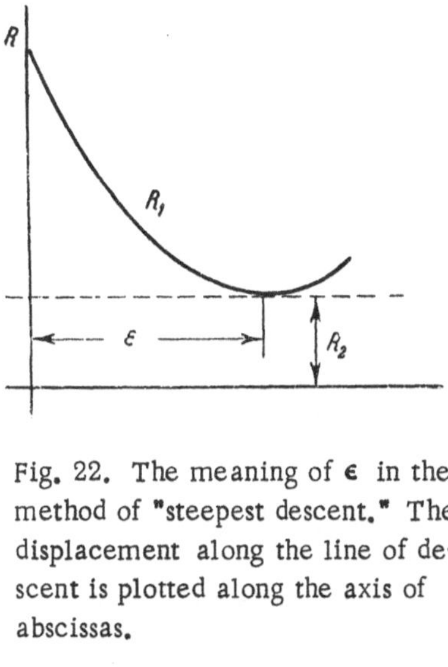

Fig. 22. The meaning of ε in the method of "steepest descent." The displacement along the line of descent is plotted along the axis of abscissas.

Thus the displacement of the point characterizing the structure along the line grad R will not follow the line of shortest descent, but will deviate from it, as is shown in Fig. 21. From point 1 we arrive at point 2. However, we can repeat the process a countless number of times. As a result we arrive at the origin of coordinates where R is equal to zero. In view of the fact that Φ_0 can not coincide exactly with Φ_c, and also because of measurement errors, it will not be possible to make R smaller than some minimum which will be denoted by R_{min}. This means that the minimum R lies on some surface near to the origin, but not at zero.

Carrying out the displacement of the representative point along 1 – 2, we discover that we have arrived at point 2 simply because further displacement would lead to an increase in R.

Thus, the appearance of R along the line 1 – 2 must be approximately such as appears on Fig. 22.

Since we know the original surface $R(x_1x_2...x_n) = R_1$, we can find the direction N of its normal. Moving along this direction, we shall arrive at a point where $\partial R/\partial N = 0$. The calculated distance along the normal from the plane $R = R_1$ to the point that was found is the shift of the representative point ϵ, and is what interests us.

The projections of the vector ϵ on the axes of coordinates give the corrections to the original coordinates.

The formulas which give the values of ϵ_i are cumbersome and impractical. Therefore we shall not discuss their derivation [55]. However, the idea of the method, clear and simple in geometric interpretation, is interesting.

5. Method of Least Squares

If Booth's method has the advantage of being geometrically graphic, the method of least squares which Hughes (1941) used for the first time in structure analysis has that of simplicity of computational formulas. Today this method is used exclusively.[1] Of course, both methods are very closely related; the goal of both is to find the minimum of (19).

Let us examine the expression

$$w(\Phi_o - \Phi_c) = w \sum_j \left(\frac{\partial \Phi_c}{\partial x_j}\right) \varepsilon_j, \tag{20}$$

which is nothing but the expansion of $(\Phi_o - \Phi_c)$ into a series limited to the first powers of the coordinate shifts ϵ_i. The number of such equations is much greater than that of the unknowns. The method of least squares gives the best solution.

Differentiating and summing over hkl, we obtain

$$\sum_{hkl} W^2 (\Phi_o - \Phi_c) \frac{\partial \Phi_c}{\partial x_i} = \varepsilon_i \sum_{hkl} W^2 \left(\frac{\partial \Phi}{\partial x_i}\right)^2 + \sum_{i'}{}' \varepsilon_{i'} \sum_{hkl} W^2 \frac{\partial \Phi_c}{\partial x_i} \frac{\partial \Phi_c}{\partial x_{i'}}. \tag{21}$$

The sum with a prime means that the term with $i = i'$ does not enter into it.

The sums in (21) may be calculated, and thus we obtain a system of equations linear with respect to the corrections to the coordinates ϵ_i.

Hughes [57] used the method in this form.

A simplified form of (21) is more widely used:

$$\varepsilon_i = \frac{\sum W^2 (\Phi_o - \Phi_c) \frac{\partial \Phi_c}{\partial x_i}}{\sum W^2 \left(\frac{\partial \Phi_c}{\partial x_i}\right)^2}. \tag{22}$$

[1]The so-called modified method of "steepest descent" [56] is actually a simplified method of least squares.

We neglect the second term in (21) on the basis of the following considerations. All the terms of the first sum in (21) are positive; the remaining sums contain terms with signs distributed at random. Thus, when there is a large field of indices, $\Sigma W^2(\partial \Phi_c/\partial x_i)(\partial \Phi_c/\partial x_i')$ must tend toward zero (it is clear that the number of indices must be much larger than the number of structural parameters). Moreover, the error of (22) in comparison with (21) is unimportant, because the process of finding the corrections ϵ_i may be repeated several times.

The method of least squares may be applied when the coordinates of all the atoms are known roughly. It is clear from the essence of the method that such atoms as are not considered in computing F_c cannot be found by minimizing R.

The atomic coordinates may be known at first with an accuracy of the order of only 0.5 A; this is sufficient to have the computed corrections ϵ_i come to a correct result.

It was shown in an article [58] that a guarantee that (22) will lead to a unique solution is present when the following expression is taken as the weight function W:

$$W^2_{hkl} = \frac{1}{f^2} d^n_{hkl},$$

where d_{hkl} is the interplanar distance, n = 4 for a two-dimensional summation, and n = 5 for a three-dimensional one.

In the process of approximating F_o by F_c, the signs of F_c must be recomputed after each stage of correction. The number of these stages may reach eight to ten in case the original structure is a crude one.

The method of least squares has an advantage over that of constructing a sequence of electron density series only in one case: all the structure amplitudes within the limiting sphere are not known. In this case an electron density series cannot be constructed, and the method of least squares will lead to an accurate structure. The application of this method should be especially recommended when the field of experimental material is reduced to data on F_{h0l} and F_{h1l} [59].

6. The Differential Method

At some stage of the analysis, let the crystal structure be known with an accuracy of the order of 0.05 A. A new set of F_c may be computed, the corresponding signs applied to F_o, and a new electron density series be summed. It will then be necessary to calculate the values of $\rho(xyz)$ at 27 points surrounding the maxima, and then, using any interpolation method [59], to find a point at which $\partial\rho/\partial x = \partial\rho/\partial y = \partial\rho/\partial z = 0$. The vectors $\mathbf{r}_j$ will then change by the small values of ϵ_j.

There is another possibility worthy of consideration from the point of view of economy of time: that of the direct calculation of the displacement vectors ϵ [60].

In fact, differentiating

$$\rho(r) = \frac{1}{V} \sum F_o \exp -2\pi i \mathbf{H}\mathbf{r},$$

we obtain

$$\frac{\partial\rho}{\partial\mathbf{r}} = \frac{1}{T} 2\pi i \sum \mathbf{H} F_o \exp 2\pi i \mathbf{H}\mathbf{r}. \tag{23}$$

The derivative at the point $(\mathbf{r}_j + \epsilon_j)$ is equated to zero. Consequently,

$$\sum \mathbf{H} F_o \exp 2\pi i (\mathbf{H}\mathbf{r}_j + \mathbf{H}\epsilon_j) = 0.$$

But, since ϵ_j is small, then $\exp 2\pi i\mathbf{H}\epsilon_j \approx 1 + 2\pi i\mathbf{H}\epsilon_j$. Thus

$$\sum \mathbf{H} F_o \exp 2\pi i \mathbf{H}\mathbf{r}_j + 2\pi i \sum (\mathbf{H}\epsilon_j) \mathbf{H} F_o \exp 2\pi i \mathbf{H}\mathbf{r}_j = 0. \tag{24}$$

It is from this equation that the vector ϵ_j may be found. For a practical computation it must be resolved into components along the axes of the reciprocal lattice.

Substituting $\mathbf{H}\epsilon = h\epsilon_x + k\epsilon_y + l\epsilon_z$ and projecting onto the $\mathbf{b}$ axis of the reciprocal lattice, we obtain

$$\begin{aligned} &\mathbf{b} \sum h F_o \exp 2\pi i \mathbf{H}\mathbf{r}_j + 2\pi \mathbf{b} i \varepsilon_x \sum h^2 F_o \exp 2\pi i \mathbf{H}\mathbf{r}_j + \\ &+ 2\pi \mathbf{b} i \varepsilon_y \sum hk F_o \exp 2\pi i \mathbf{H}\mathbf{r}_j + 2\pi \mathbf{b} i \varepsilon_z \sum hl F_o \exp 2\pi i \mathbf{H}\mathbf{r}_j = 0 \end{aligned} \tag{25}$$

and two more analogous equations—projections onto the other two axes.

Thus, the three unknowns constituting ϵ_j can be determined by calculating the values of twelve sums at the point $\mathbf{r}_j$.

The method may be simplified, if the spherical symmetry of the electron density distribution near the atomic center is taken into account. This assumption is very near the truth.

It is not hard to see that (23a) may be rewritten in the form

$$\frac{\partial \rho}{\partial x}+\varepsilon_x \frac{\partial^2 \rho}{\partial x^2}+\varepsilon_y \frac{\partial^2 \rho}{\partial x \partial y}+\varepsilon_z \frac{\partial^2 \rho}{\partial x \partial z}=0 \qquad (26)$$

together with two analogous equations. When the atom has spherical symmetry and has its center at $x_0y_0z_0$, the electron density depends only on the distance of the point xyz from $x_0y_0z_0$. Denoting the square of this distance by g, we have, for example, for monoclinic symmetry

$$g=(x-x_0)^2+(y-y_0)^2+(z-z_0^2)+2(x-x_0)(z-z_0)\cos\beta .$$

Thus, derivatives with respect to the coordinates are exchanged for those with respect to g, namely,

$$\frac{\partial^2 \rho}{\partial x^2}=\frac{\partial^2 \rho}{\partial y^2}=\frac{\partial^2 \rho}{\partial z^2}=2\frac{\partial \rho}{\partial g},$$

but

$$\frac{\partial^2 \rho}{\partial z}=2\frac{\partial \rho}{\partial g}\cos\beta,$$

whence, as can be shown,

$$\varepsilon_x=\lambda_l \cos\beta-\frac{\lambda_h}{\sin^2\beta}, \quad \varepsilon_y=-\lambda_k, \quad \varepsilon_z=\lambda_h\cos\beta-\frac{\lambda_l}{\sin^2\beta},$$

where

$$\lambda_h=\frac{\dfrac{\partial \rho}{\partial x}}{2\dfrac{\partial \rho}{\partial g}}$$

etc. Thus, to compute the vector ϵ the sum of four series at the point $x_0y_0z_0$ must be computed: $\partial\rho/\partial x$, $\partial\rho/\partial y$, $\partial\rho/\partial z$ and any of the second derivatives instead of $\partial\rho/\partial g$.

In another article [65] a variation of the same method appears, possibly of interest for such rarely studied crystals that are without a center of inversion.[2] In this second variation, account is taken of the continuous change in phase angles which occurs when an atom is shifted along the vector ϵ.

[2] Translator's note. Such crystals are now studied very frequently.

7. The Method of Difference Series

This method is used primarily for finding the best atomic coordinates. Furthermore, if the measurements have been made with high accuracy, it is the most suitable way to refine the values of the electron density, i.e., to come to conclusions about the number of electrons in the atom, about the anisotropy of the atomic function, and, consequently, about the temperature factor. Moreover, this series is convenient for showing the presence of light atoms in the presence of heavy ones.

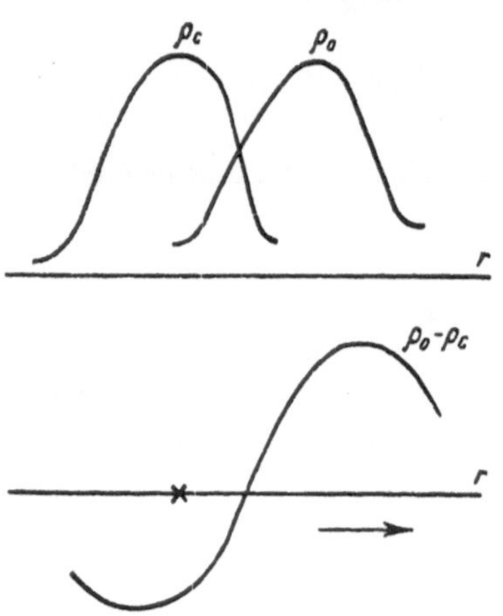

Fig. 23. Method of difference series.

$$\rho_o = \sum F_o \exp 2\pi i \mathbf{Hr}$$

and

$$\rho_c = \sum F_c \exp 2\pi i \mathbf{Hr} \qquad (27)$$

Let us assume that the centers of some atoms in the series are somewhat shifted. It is evident (Fig. 23), that by subtracting ρ_c from ρ_o a distribution is formed consisting of maxima and minima. The atomic coordinates used in calculating F_c should be shifted in the direction of the maxima of the difference series. The magnitude of this shift is easily calculated in the following manner.

Let us approximate an atom near its maximum by any suitable function, for example,

$$\rho(r) = \Delta(1 - pr^2). \qquad (28)$$

Let $\rho_o(r)$ and $\rho_c(r)$ be shifted with respect to each other by the magnitude Δ. Assuming that the constants A and p of these two functions are about the same, we obtain

$$D(r) = \rho_o(r) - \rho_c(r) = A\,[-(1 - pr^2) + (1 - pr^2 + 2pr\Delta - p\Delta^2)],$$

whence,

$$\frac{dD}{dr} = 2Ap\Delta. \qquad (29)$$

Thus, the shift Δ can be calculated. The value of dD/dr is taken at the point having the atomic coordinates used in calculating F_c. It is then possible to recalculate the F_c values, and, if necessary, to change their signs, and to repeat the procedure.

Light atoms in the presence of heavy ones may appear in electron density series not as maxima, but only as a distortion of the atomic function of the adjacent heavier atom. If in this case a $\rho_O - \rho_C$ series be constructed where ρ_C is composed of the heavier atoms, then the light atoms will reveal themselves as separate maxima ([61] and [62]). Further refinement of the structure may be continued by repeating the procedure.

We assume that the task of refining the electron density pattern, which has for its goal not only the finding of the coordinates of the maxima of the series, but also the analyzing of the number of electrons, of the anisotropy of the electron cloud, and of the other characteristics of the atomic functions, is among the most complicated ones and lies on the boundary of the accuracy of the method. Problems of this type may be solved only in those cases for which the accuracy of determining F_O experimentally is no worse than 2-5%. It is difficult to attain this result for a number of reasons. The most important are: 1) the great sensitivity of the absorption factor to the sizes and shapes of the samples, 2) the difficulty in accounting for the primary and secondary extinction, 3) nonuniqueness of the separation of the intensity scattered by the mean electron density from the background and 4) lattice distortions.

If the researcher is sure that all the factors are taken into consideration, then he may attempt to continue the approximation of F_O by F_C using the method of difference series and taking into account the anisotropy of the atomic functions.

This method has been discussed in Cochran's articles cited above. We do not consider that the above indicated accuracy in the F values can be attained, and therefore we assume articles with this tendency to be devoid of interest [54].[3]

8. Accuracy Factors

The result of a structure analysis is an electron density series

$$\rho(r) = \frac{1}{V} \sum^{H_0} F_{\mathbf{H}} \exp -2\pi i \mathbf{Hr}. \tag{30}$$

Since at large $\mathbf{H}$ the reflection intensities begin to merge with the background, the series $\rho(\mathbf{r})$ will inevitably be cut off at some value H_0. Thus, the first factor affecting the accuracy of $\rho(\mathbf{r})$ is this severance of the series. The magnitudes $F_{\mathbf{H}}$ are calculated from the intensities. The intensities themselves are measured with only a limited accuracy.

[3]Translator's note. Modern measurements are frequently made with precisions of up to 2-3% in F and the difference series method is in frequent use outside the USSR.

Furthermore (see chapter I) there is some indefiniteness in breaking up the intensities into the part dependent on the mean electron density (which we determine), and the part of the scattering depending on the correlation of the thermal or other disorder in the lattice. There are great difficulties in taking account of extinction and absorption. In the usual experiment, due to all these reasons taken together, the F values are known with an accuracy of the order of several tens of percent. In one way or another, the second factor affecting accuracy is the precision of our knowledge of the structure amplitudes.

There will be no grievous error if we assume [63] that the deviations of F_0 from the true values (ΔF) are distributed according to the Gaussian law. Let us denote the relative mean square deviation by b:

$$b^2 = \overline{\left(\frac{\Delta F}{F}\right)^2}. \tag{31}$$

Thus the value of the residual terms of the series and the magnitude b determine the accuracy of our knowledge of the electron density $\rho(\mathbf{r})$ and its derivatives.

The error in the value of $\rho(\mathbf{r})$ is evidently equal to

$$\frac{1}{V}\sum^{H_0} \Delta F \exp - 2\pi i \mathbf{H}\mathbf{r} + \sum{}', \tag{32}$$

where Σ' is the residual term.

Indeed, for various values of $\mathbf{r}$ the errors in determining ρ are different. Assuming that the errors depending on $\mathbf{r}$ are also distributed according to the Gaussian law, then, raising to the square and averaging, we shall find the mean square error of the electron density (average over the cell)

$$\overline{(\Delta\rho^2)} = \frac{b^2}{V^2}\sum F^2_{\mathbf{H}} + \frac{1}{V^2}\sum{}' F^2. \tag{33}$$

It is essential, however, to recall that the influence of the residual term is systematic.

Actually, the electron density of the atomic structure may be written in the form

$$\rho = \frac{1}{V}\sum_{H}\sum_{k=1}^{N} f_k \exp - 2\pi i \mathbf{H}(\mathbf{r} - \mathbf{r}_k),$$

i.e.,

$$\rho = \sum_{k=1}^{N} \rho(\mathbf{r} - \mathbf{r}_k), \tag{34}$$

where $\rho(\mathbf{r} - \mathbf{r}_k) = \frac{1}{V} \sum_{N} f_k \exp 2\pi i \mathrm{H}(\mathbf{r} - \mathbf{r}_k)$ is the atomic function of the _k_th atom.

Thus, the residual term Σ' of the series may be considered to be the result of the superposition of the waves due to the break off of the N atomic functions.

The interference of these waves cannot be taken into account and must be considered as random in every respect except one – a sharp decrease of the maximum of the atomic function.

This systematic influence of the break-off waves is concentrated within a small volume with a radius of the order of 0.5 A and, averaging over the entire volume of the cell, it contributes very little to Σ'. Therefore, the systematic character indicated above of the influence of the break-off wave does not affect the accuracy of determining the electron density far from the maximum; the random interference of the break-off waves is the determining factor. It is different at the positions of the maxima. Here $\underline{\rho}$ ($\mathbf{r}$) leads to sharply lowered density values and the formula given for $\overline{(\Delta\rho)^2}$ will give, not the deviation of the value of ρ_0 from the true one, but only the probable scatter from fictitious lowered atomic densities of the same kind. It is with this in mind, that we applied the computation of $\overline{(\Delta\rho)^2}$ to tetraiodoethylene [59].

To find values of the electron densities closest to the true ones the series must be continued to the theoretical limit at infinity.

One of the methods is to construct the series

$$\rho_\infty = \rho_0 + \sum_{H_0}^{\infty}{}' F_c \exp 2\pi i \mathbf{Hr}, \tag{35}$$

i.e., to substitute the amplitudes F_c for F_0 in the residual term.

Unfortunately it is difficult to establish the accuracy of the evaluation of ρ_∞, since we do not know how greatly the F_c differ from the true values of the structure amplitudes. It can only be said that this deviation is at any rate greater than _b_, and the limiting accuracy of the value of the true density can be evaluated by the formula

$$\overline{(\Delta\rho^2)} = \frac{b^2}{V^2}\left(\sum^{H_0} F_o^2 + \sum^{\infty}{}' F_c^2\right). \qquad (36)$$

The other method is to introduce a fictitious temperature factor (a mental increase in the temperature of the experiment). Furthermore, the temperature is "increased" in such a way that the series converges within the limiting sphere.

There is no difference between these two methods. The accuracy with which the electron density can be evaluated is represented by the same formula, i.e.,

$$\overline{(\Delta\rho^2)} = \frac{b^2}{V^2}\sum F_T^2, \qquad (37)$$

where F_T are the structure amplitudes at the "increased" temperature.

9. Computation of the Errors in the Determination of Atomic Coordinates [63]

We define an atomic center as the position of a maximum in the electron density. Therefore, the error in an atomic coordinate can be found from the formula

$$\Delta r = \frac{\Delta\left(\frac{\partial\rho}{\partial r}\right)}{\frac{\partial^2\rho}{\partial r^2}}.$$

The value of the second derivative at the maximum is easily found from the experimental series. As has already been noted, within a radius of about 0.5 A the maximum can be easily approximated by the expression

$$\rho = \alpha - \beta r^2, \qquad (38)$$

where α and β are constants. It is clear that $\partial^2\rho/\partial r^2 = -2\beta$.

For example, for organic crystals the value of β is close to $\beta = 50$ el/A^5.

As concerns $\Delta(\partial\rho/\partial r)$, its value is computed analogously to that of $\Delta\rho$.

Then

$$\frac{\partial \rho}{\partial r} = \frac{1}{V} \sum 2\pi i H F \exp -2\pi i \mathbf{H}\mathbf{r}, \tag{39}$$

$$\Delta\left(\frac{\partial \rho}{\partial r}\right) = \frac{1}{V} \sum 2\pi i H \exp -2\pi i \mathbf{H}\mathbf{r} \Delta F + \sum{}', \tag{40}$$

where Σ' is the residual term of the series.

Squaring and averaging, we obtain

$$\overline{\left[\Delta\left(\frac{\partial \rho}{\partial r}\right)\right]^2} = \frac{1}{V^2} \sum (2\pi H)^2 \overline{(\Delta F)^2} + \frac{1}{V^2} \sum{}' (2\pi H)^2 F^2 \tag{41}$$

or

$$\overline{\left[\Delta\left(\frac{\partial \rho}{\partial r}\right)\right]^2} = \frac{1}{V^2}\left[b^2 \sum (2\pi H)^2 F^2 + \sum{}' (2\pi H)^2 F^2\right]. \tag{42}$$

Computation from this formula may be done quickly and easily by breaking up the sphere of reflection into five to ten shells. The value of Σ' is evaluated from theoretical formulas (see below).

10. An Approximate Evaluation of the Sums That Enter into the Error Formulas

In the formulas of the preceding paragraph, there are primed sums with values unknown from experiment. These sums may be evaluated by substituting F^2 by its average value, equal to Σf_j^2 as we know, where f is the atomic factor of the jth atom.

If we also evaluate the sums taken from the measured reflections in this manner, we shall obtain a priori formulas for the errors in determining the electron density and the atomic coordinates. Experiment shows that ΣF^2, and $\Sigma H^2 F^2$, and similar ones constructed for real structures from a sufficiently large number of measured reflections differ very little from their mathematical expectations. Such a procedure is justified in evaluating the errors when we are interested in the order of magnitude only.

The calculation consists in substituting Σf_j^2 for F^2 and then in using the relationship $f_j = \hat{f} Z_j$, where $\hat{f}(H)$ is the unitary atomic curve. Integration may be substituted for the calculation of the sums.

Thus the evaluation of the error[3] is carried out by the substitution of $\Sigma Z_j^2 \int \hat{f}^2 d\tau$ for ΣF^2, where $d\tau$ is the element of volume of reciprocal space. The formula for the a priori error in the electron density, under the conditions that the f curves are similar, becomes

$$\overline{(\Delta\rho)^2} = \frac{b^2}{V} \sum Z_j^2 \int_0^{\tau} \hat{f}^2 d\tau + \frac{1}{V} \sum Z_j^2 \int_{\tau}^{\infty} \hat{f}^2 d\tau. \tag{43}$$

the V is introduced when the integral is substituted for the sum.

An analogous formula for the error in the derivative of the density along a direction has the form

$$\overline{\left[\Delta\left(\frac{\partial\rho}{\partial r}\right)\right]^2} = \frac{\Sigma Z_j^2}{V}\left[b^2 \int_0^{\tau} (2\pi H)^2 \hat{f}^2 d\tau + \int_{\tau}^{\infty} (2\pi H)^2 \hat{f} d\tau\right]. \tag{44}$$

The integrals can be evaluated either graphically or by using some simple functional approximation to the unitary atomic curve.

The results of the computation are as follows: If the series converges, i.e., when S can be considered as equal to infinity, then

$$\Delta r = 0.5\alpha^{\frac{5}{2}} b \sqrt{\sum \frac{1}{V}\left(\frac{Z_i}{Z_0}\right)^2}, \tag{45}$$

where α is the constant in the formula $f = ke^{-\alpha s}$ selected for the approximation of the atomic curve.

As we have noted above, a series can always be made to converge at the expense of an increase in α. The normal value of α is 0.35, and a value at which a series may be considered convergent for copper radiation is $\alpha = 1$. As a concrete example let us take an organic structure. Then $Z_i/Z_0 = 1$ and the number of atoms per unit volume n/V is quite constant for aromatic hydrocarbons and equal to about 0.05 A^{-3}. Thus, $\Delta r = 0.08b\alpha^{5/2}$. In case copper radiation is used, the series becomes convergent when $\alpha = 1$ and, consequently, with an accuracy of measurement b = 0.1, the three-dimensional series makes it possible to find interatomic distances with an accuracy of up to 0.01 A.

Let us see now whether it would not be more advantageous not to introduce artificial convergence of the series, and to work on a series broken off at the copper radiation limit, but now with $\alpha = 0.35$.

[3]See page 261. See also [4].

For $\alpha = 0.35$ and $S = 8$, we obtain

$$\Delta r = 0.06 \cdot \sqrt{0.01 + 0.015b^2} \cdot \tag{46}$$

With b = 0, the magnitude of $\Delta r = 0.006$. This is the limit of accuracy. With b = 0.1 we obtain practically the same magnitude. With a rough estimate (b = 1) we obtain $\Delta r = 0.01$ A. Thus we come to an important conclusion: In determining the coordinates of the maxima of the series there is no sense in introducing an artificial convergence. It would have meaning in an experimental unrealizable case – for example in work accurate to b = 0.01.

The influence of errors in determining F on the accuracy of the coordinates is also very interesting. The following table can be constructed for the example under study.

b....	0	0.1	0.5	1 0	1.5	2.0	5.0	10.0
Δr...	0.006	0.006	0.007	0.01	0.013	0.02	0.04	0.08

Thus, with an average error in F of one order of magnitude we still obtain an accuracy of up to 0.1 A. This calculation clarifies Lukesh's [64] computational results; he showed that by dividing all the spots of an x-ray pattern into three groups according to their intensities, it is possible to obtain a roughly correct (with respect to the distribution of the maxima) picture of the electron density.

We should also dwell on another special point, namely, the proportionality of Δr to the quantity

$$\sqrt{\sum_{i=1}^{q} \left(\frac{Z_i}{Z_0}\right)^2}.$$

Here it is interesting to discuss the mutual influence of atoms of different atomic numbers in the crystal under study.

Let us compare, for example, the accuracy of determining a carbon atom in isomorphous structures which have one heavy atom per five light ones, namely an atom of chlorine, bromine, or iodine. In the first case the error is proportional to

$$\sqrt{5 + \left(\frac{17}{6}\right)^2} = 3.6,$$

in the second case

$$\sqrt{5+\left(\frac{35}{6}\right)^2}=6.4$$

and in the third case

$$\sqrt{5+\left(\frac{53}{6}\right)^2}=9.1.$$

The incorporation of heavy atoms into a molecule is widely used as a method of structure analysis, as is well known. Now we see "the reverse side of the coin." Introducing an atom of chlorine increases the error in determining the carbon by 1.5 times, introducing bromine by 2.5 times, and introducing iodine by more than four times.

It is true that the determination of "heavy" atoms in such structures can be done very accurately (the heavier the atom the better).

In view of the wide application of the projection method in structure analysis, let us analyze the formula for the error in determining the coordinates of the maxima of the series σ (r).

In the most frequent cases (α = 0.35 and α = 8) obtained by means of analogous calculations, we get

$$\Delta\left(\frac{\partial\sigma}{\partial r}\right)^2=(k)^2\frac{\sum Z_j^2}{4\pi S}(20b^2+5), \tag{47}$$

where S is the area of the projection.

In the case of a broken off series it is impossible to express the second derivative of σ (r) by a simple projection. When there is no break, the value of the second derivative is easily calculated. It is equal to

$$\left[\frac{\partial^2\sigma}{\partial a}\right]_{a=0}=-\frac{3k}{2\pi a^4}, \tag{48}$$

i.e., with α = 0.35, it is equal to $-32k$. By analogy with the three-dimensional series we can ascertain that a break will decrease this magnitude at least twofold. Thus, for a projection series under the given conditions,

$$\Delta r^2=\frac{1}{4\pi(16)^2}\sum\frac{n}{S}\left(\frac{Z_i}{Z_0}\right)^2(20b^2+5). \tag{49}$$

In a structure made up of one kind of atoms we obtain $n/S = 0.2$, and for a flat aromatic molecule

$$\Delta r = 0.04\sqrt{b^2 + 0.25}. \tag{50}$$

We see that the ultimate accuracy of determining atomic coordinates by an electron density projection is 0.02 A when the series is terminated by using CuK_α.

Introducing artificial convergence of the series is not recommended any more than in the three-dimensional case. As has been indicated above, $\alpha = 1$ for this purpose, and S can be assumed equal to ∞. Under these conditions $\Delta r = 0.16$. Consequently, only with very accurate work ($b = 0.1$) will results be obtained which are no worse than those given by the "natural" series.

It is interesting, likewise, to give a table here which shows the capability of rough measurements, when working with a terminated series. For the given example we obtain,

b....	0	0.1	0.5	1.0	1.5	2.0	5.0	10.0
Δr...	0.02	0.02	0.028	0.04	0.06	0.08	0.2	0.4

We see that the deterioration in the accuracy of structure determination due to measuring amplitudes to an accuracy of 50% (that means intensities to 100%) is reflected only slightly in the results of the analysis. Further increase of Δr with increase in experimental error occurs in the projections $\sigma(r)$ much more rapidly than in the three-dimensional series.

CONCLUSION

The theory of structure analysis in the sense that has been used throughout this book has progressed considerably during these recent years. The computing machines that are being used for structure determination by diffraction methods will soon make possible the solution of a series of problems of method which as yet remain unsolved. The task of finding the structure may be assigned to the machine.

Not every structure can be solved automatically. The magnitude $\sigma = (\Sigma n_j^2)^{\frac{1}{2}}$ is the basic criterion for determining the solvability of a structural problem. If the value of $\sigma > 0.2$ then a large number of reliable connections exist between the signs of the structure amplitudes. By using the simplest connecting determinants (m = 3,4) the signs of a considerable fraction of the amplitudes can be determined, and a first approximate series of electron density can be constructed. The structure may then be defined more accurately by any of the methods analyzed in chapter VI.

If $0.15 < \sigma < 0.2$, then the number of signs determined with certainty will be sufficient so that, by using them as basic, many other amplitudes can be found with the help of statistical considerations, that is, basing ourselves on the preponderant positivity of the structure product. The further course of the analysis will be the same.

In case $0.1 < \sigma < 0.15$, the solution of the structure problem becomes problematic. Here it is necessary to pursue two paths: To apply an analysis of the convolution of the electron density and simultaneously to attempt the determination of the signs of the amplitudes by a statistical method. The application of connecting determinants of high orders is quite possible for finding reliable relations between signs. Only for structures of this degree of complexity do the problems take on a creative character: perhaps the carrying out of piecemeal tests, and the application of geometric analysis, etc., may be necessary.

For $\sigma < 0.1$ the solution of the structure problem by direct methods becomes impossible. For the most complicated structures we can, with painstaking work, test the coincidence of different structural models with the experimental data, and ascertain merely that the model does not contradict the experiment.

Of course, the correctness of the structure of a molecule is very important (for example, large aromatic molecules). In this case the limits of solvability converge. The reason is obvious: A series of peaks coincides in the convolution of the electron density. For the method of direct sign determination this carries with it the possibility of successfully constructing connecting determinants of high order, and also of distorting the Gaussian distribution of the amplitudes, so that among them, will occur some of absolute magnitude $4-5\sigma$.

The question concerning the comparative merits of constructing a convolution or of applying the method of direct sign determination then becomes practically important. Probably (this is exactly one of the problems of method which must be solved by computing machines) the possibilities of both methods are approximately equal – where one is possible, so is the other. When the convolution cannot be interpreted, the signs cannot be found by direct methods.

However, the sign method has the decided merit of being easily carried out on a machine. The construction of structure products, their comparison, and the sign determination of amplitudes may be made automatic. It is hardly possible to interpret automatically a three-dimensional series of the convolution of the electron density. The sign method is also probably advantageous if the structure is composed of similar atoms. In the case of crystals without a center of inversion the Patterson series is still without peer. We have said very little in this book about direct methods of phase determination for crystals without a center of symmetry. However, the general character of the theory points surely to the possibility of a practical solution for this problem too.

The basic task of structure analysis is the finding of a rough structure. It is always possible to refine it more accurately by machine, and the interatomic distances can then be found with an accuracy basically dependent on the volume of the experimental material. Interatomic distances determined with an accuracy of up to 0.01-0.05 A is the sole result of a structure analysis based on the study of the intensity of scattering of the mean square electron density.

The anisotropy of the shape of the atom is clearly perceptible in simple structures on maps of the electron density. However it is hardly expedient to study, even in slightly complicated cases, the anisotropic thermal vibrations by the method of x-ray structure analysis. The method is not sufficiently accurate for this.

The problems of determining the degrees of ionization of the atoms, and also those connected with the analysis of the behavior of valence electrons, must be solved by other physical methods. Possibly the study of the intensity of scattering from the correlations between the deviations of the electron distribution from the mean are basically important in solving these problems. At present, however, the more promising methods are those having nothing in common with diffraction phenomena, and the most promising of all are those in which nuclear resonance is studied by spectroscopic methods.

Structure analysis based on a study of the amplitudes of scattering from the mean square electron density preserves its value primarily as a method for finding the geometric configuration of a molecule. The area of application of this method includes a huge number of complicated molecules containing 15-20 atoms, and even more if a heavy atom is present. The automation of structure analysis will allow it to compete successfully in many cases with chemical methods of determining the stereochemical configuration of a molecule.

The progress of recent years in structure analysis has brought complete clarification, and has determined the possibilities and the value of this important method for chemistry, mineralogy, and metallography.

BIBLIOGRAPHY

[1] R. Hosemann and S. N. Bagchi, Acta Cryst., 5, 749 (1952).

[2] R. Hosemann and S. N. Bagchi, Acta Cryst., 6, 318 (1953).

[3] M. A. Blokhin, "Physics of X-Rays" (Gostekhizdat, 1953).).

[4] A. I. Kitaigorodskii, "X-Ray Structure Analysis," (Gostekhizdat, 1950).

[5] R. W. James, "Optical Principles of X-Ray Diffraction," (Foreign Literature Press, 1953).

[6] R. McWeeny, Acta Cryst., 4, 513 (1951).

[7] R. McWeeny, Acta Cryst., 5, 463 (1952).

[8] R. McWeeny, Acta Cryst., 6, 631 (1953).

[9] A. L. Patterson, Phys. Rev., 56, 973 (1939).

[10] P. P. Ewald, Proc. Phys. Soc., 52, 167 (1940).

[11] D. Harker and J. S. Kasper, Acta Cryst., 1, 56 (1948).

[12] A. J. Kitaigorodskii, Doklady Akad. Nauk, USSR, 101, 1 (1955).

[13] B. K. Vainshtein and Z. G. Pinsker, Doklady Akad. Nauk, USSR, 64, 49 (1949).

[14] B. K. Vainshtein, "Structural Electron Diffraction," (Akad. Nauk USSR, (1956).

[15] G. E. Bacon and R. S. Pease, Proc. Roy. Soc., 220, 397 (1953).

[16] A. C. J. Wilson, Acta Cryst., 2, 317 (1949).

[17] A. C. J. Wilson, Nature, 150, 152 (1942).

[18] Bernshtein, "Theory of Probabilities," (Izd. Akad. Nauk, USSR).

[19] Karle and Hauptman, Acta Cryst., 6, 131 (1953).

[20] A. I. Kitaigorodskii, Doklady Akad. Nauk, 94, 2 (1954).

[21] A. I. Kitaigorodskii, Trudy Inst. Krist., 10 (1954).

[22] A. J. C. Wilson, Acta Cryst., 3, 258 (1950).

[23] A. I. Kitaigorodskii, Zhur. Fiz. Khim. 25, 127 (1951).

[24] A. J. C. Wilson, Zhur. Fiz. Khim., No. 10 (1953).

[25] A. J. C. Wilson, Research, 4, 141 (1952).

[26] A. J. C. Wilson, Nuovo cimento, 9, 1 (1952).

[27] D. Rogers and A. J. C. Wilson, Acta Cryst., 6, 439 (1953).

[28] E. R. Howels, D. C. Phillips, and D. Rodgers, Acta Cryst., 3, 211(1950).

[29] E. W. Hughes, Acta Cryst., 2, 34 (1949).

[30] D. Harker, Acta Cryst., 6, 731 (1953).

[31] G. Kartha, Acta Cryst., 6, 813 (1953).

[32] B. K. Vainshtein, Doklady Akad. Nauk, USSR, 93, 821 (1953).

[33] D. Rodgers, Acta Cryst., 3, 455 (1950).

[34] A. I. Kitaigorodskii, Zhur. Fiz. Khim., 24, 747 (1953).

[35] W. H. Zachariasen, Acta Cryst., 5, 68 (1952).

[36] D. Sayre, Acta Cryst., 5, 60 (1952).

[37] Bertaut and Pepinsky, Acta Cryst., 7, 214 (1954).

[38] Lavigne, Acta Cryst., 5, 846 (1952).

[39] Gillis, Acta Cryst., 1, 76 (1948).

[40] A. I. Kitaigorodskii, Doklady Akad. Nauk., 105, 482 (1955).

[41] V. I. Smirnov, Course of Higher Mathematics (Gostekhizdat, 1949).

[42] Goedkoop, Acta Cryst., 3, 374 (1950).

[43] A. I. Kitaigorodskii, Doklady Akad. Nauk., USSR, 102, 519 (1955).

[44] Rumanova, Trudy Inst. Cryst., No. 10, 59 (1954).

[45] M. Buerger, Acta Cryst., 3, 87 and 465 (1950).

[46] A. I. Kitaigorodskii, Zhur. Ékspt. i Teor. Fiz. 21, 717 (1951).

[47] V. V. Sanadze and G. I. Zhdanov, Doklady Akad. Nauk, USSR, 73, 111 (1950).

[48] M. Buerger, Acta Cryst., 4, 531 (1951).

[49] B. K. Vainshtein, Doklady Akad. Nauk USSR, 78, 1137 (1951).

[50] K. Tauer and W. N. Lipscomb, Acta Cryst., 5, 606 (1952).

[51] M. Atoji and W. N. Lipscomb, Acta Cryst., 7, 173 (1954).

[52] A. J. C. Wilson, Acta. Cryst., 3, 397 (1950).

[53] W. C. Hamilton, Acta Cryst., 8, 185 (1955).

[54] A. I. Kitaigorodskii, Nature, 179, 410 (1957).

[55] A. D. Booth, Nature, 160, 196 (1947).

[56] Qurashi, Acta Cryst., 2, 404 (1949).

[57] E. W. Hughes, J. Am. Chem. Soc., 63, 1737 (1941).

[58] Qurashi, Acta Cryst., 6, 577 (1953).

[59] A. I. Kitaigorodskii, T. L. Khotsianova, and Yu. T. Struchkon, Zhur. Fiz. Khim., 27, 1490 (1953).

[60] A. D. Booth, Trans. Far. Soc., 42, 444 (1946).

[61] W. Cochran, Acta Cryst., 4, 81 (1951).

[62] W. Cochran, Acta Cryst., 6, 260 (1953).

[63] A. I. Kitaigorodskii, Zhur. Fiz. Khim. 20, 397 (1950).

[64] J. Lukesh, J. Appl. Phys., 18 (1947).

[65] A. D. Booth, Trans. Far. Soc., 42, 617 (1946).